A PRACTICAL GUIDE for

Hazardous Waste Management, Administration, and Compliance

James L. Lieberman

Edited by
Gary Gautier

LEWIS PUBLISHERS
Boca Raton Ann Arbor London Tokyo

Library of Congress Cataloging-in-Publication Information

Lieberman, James L.
A Practical Guide for Hazardous Waste Management, Administration, and Compliance / by James L. Lieberman : Gary Gautier , editor.
p. cm.
Includes bibliographical references and index.
1. Hazardous waste--Management. 2. Hazardous waste--Law and legislation--United States. I. Gautier, Gary. II. Title.
TD1030.L54 1994
363.72 ' 8705--dc20 94-12877
ISBN 1-56670-115-5

Direct all inquiries to CRC Press, Inc., 2000 Corporate Blvd., N.W., Boca Raton, Florida 33431.

Lewis Publishers is an imprint of CRC Press

International Standard Book Number 1-56670-115-5

Library of Congress Card Number 94-12877

Printed in the United States of America
1 2 3 4 5 6 7 8 9 0

Printed on acid-free paper

To my wife Leanne for her patience and support, and to my son Ben.

PREFACE

This book was written to provide clear and concise guidance for establishing a hazardous waste management, administration and compliance program that is economical, reduces present and future liability, and fulfills applicable regulations. You will find that the author chooses to take a middle of the road approach, balancing liability reduction against economic costs.

This book will not address every situation; it will, however, provide the reader with a solid understanding of a logical hazardous waste compliance program and provide the knowledge needed to select consultants for specialized projects. There are three different regulatory categories for hazardous waste generators: large quantity, small, and conditionally exempt small quantity. Generators that are large and small quantity generators (i.e., >1000 kg/mo. and 100 to 1000 kg/mo.) will find this book most useful. [Small quantity and conditionally exempt small quantity generators may also consult Thomas P. Balf, *How to Comply with Hazardous Waste Laws* (Elgin, IL: S-K Publishing, 1991) and Russell W. Phifer and William R. McTigue, Jr., Handbook of Hazardous Waste Management for Small Quantity Generators (Chelsea, MI: Lewis Publishers, 1988)].

This presentation will follow the logical steps necessary to establish a Resource Conservation and Recovery Act (RCRA) Waste Management Program. A reader unfamiliar with RCRA is encouraged to read the book in the order presented. More experienced readers will find it a helpful reference.

The regulations concerning hazardous waste management will not be reproduced in this book. Those regulations can be found in sections 29, 40 and 49 of the Code of Federal Regulations (CFR). Regulations are updated yearly and available from the Government Printing Office. Regulatory citations will be noted where appropriate.

The reader will find helpful additional subject area information sources at the end of the book. Phone numbers for general and specific information are listed in appendices.

ABOUT THE AUTHOR

James L. Lieberman CIH, has over 17 years of experience in the area of waste management. As president of Environmental Information Services, Incorporated (EIS), he directly supervises hazardous and radioactive waste sampling, analysis, characterization, profiling, transportation and disposal. In addition, Mr. Lieberman assists EIS clients with initial permitting and modifications, compliance reports and audits, and waste management programs. Mr. Lieberman formerly worked for NFT, Incorporated, a firm specializing in serving the Department of Energy Complex with hazardous and mixed waste support and compliance programs. He has conducted radioactive waste characterization and stabilization projects for Department of Energy facilities. His prior industrial experience includes the operation of a wastewater treatment plant and water works for Vail Associates, Inc., where his responsibilities included the permitting of the wastewater treatment plant and NPDES compliance. Mr. Lieberman currently holds a design patent for a portable water filter. He received his BS degree in chemistry from the University of Richmond, and an MBA with emphasis in finance from the University of Colorado. He holds a B/B license from the State of Colorado for Water and Wastewater Operation, and was certified by the American Board of Industrial Hygiene for the Comprehensive Practice of Industrial Hygiene. He has taught waste management and safety courses to both private and government groups, and was honored by Colorado Mountain College with an award for Instructor of the Year.

ABOUT THE EDITOR

Gary Gautier PhD advises Mr. Lieberman and EIS in all phases of technical writing, editing, proofreading, and correspondence. He received a BA from the University of Southwestern Louisiana, MA from the University of Texas, and PhD from the University of Colorado. He has taught a variety of courses in literary and cultural analysis, and in expository writing, for the University of Colorado's English, Humanities and Honors Departments, as well as for the University Writing Program.

TABLE OF CONTENTS Page

ACKNOWLEDGMENTS

The author would like to acknowledge the assistance of Fred Linton, Environmental Control Manager for Coors Brewing Company. A significant portion of the waste profiling system was designed by Mr. Linton. I would also like to acknowledge Vaughn Stickdorn, Industrial Waste Specialist for Coors Brewing Company, for his practical suggestions and recommendations for hazardous waste handling. We would also like to thank Holland and Hart, Attorneys at Law, for permission to reprint sections of the document entitled *Polychlorinated Biphenyls (PCB) Regulations.*

A significant portion of this book and material is edited, publicly available information. One document, the draft RCRA Orientation Manual, 1990, edition, USEPA, was used extensively.

I would also like to thank the following peer reviewers for their helpful comments and corrections. Special thanks to David M. Packard, Attorney at Law with Hutchinson, Black and Cook, for his detailed review, editing, and suggestions. Thanks to Steven A. Franklin, J.B., for his review of citations, and. Denise Gelston for her review of regulatory interpretation; both work for S.M. Stoller Corporation.

I would also like to thank Ben Phelan and Ed Perkins of Environmental Information Services for their diligent typing, formatting, and proofreading of the text.

CHAPTER 1

REGULATORY OVERVIEW AND BACKGROUND

This book presents a step-by-step approach to the actual tasks and programs that must be initiated to create a hazardous waste management program. The individual charged with initiating and maintaining a compliance program, referred to in this book as the RCRA Administrator, cannot effect his compliance program without an understanding of the statutory laws and regulations that have evolved to regulate hazardous waste, and of general environmental regulations. He must be aware of the Acts of Congress, the Resource Conservation and Recovery Act, the Hazardous Materials Transportation Act, and the Occupational Safety and Health Act, which directly affect the scope and framework of a compliance program. He should be aware of the general requirements of these laws and the regulations written and promulgated by the different agencies. The promulgated regulations are printed in the Federal Register (FR) and codified once a year into the Code of Federal Regulations (CFR).

A significant body of law exists and is constantly expanding that will impact, directly or tangentially, hazardous waste generation, storage, transportation, treatment and disposal facilities. The RCRA Administrator's operations, in addition to those mentioned above, will most likely be affected directly by the Comprehensive Environmental Response, Compensation and Liability Act (CERCLA), or Superfund, and tangentially by the Clean Air Act Amendments of 1990.

If the RCRA Administrator is tasked with the larger responsibilities of a facility Environmental Compliance Administrator, he must acquaint himself with the larger body of environmental regulations. The following regulatory overview and background will help provide some background on the acts or federal laws that will impact his program.

CONGRESSIONAL ENVIRONMENTAL ACTS

Congress has created a broad federal regulatory scheme governing hazardous wastes and substances. The Resource Conservation and Recovery Act (RCRA) profiles cradle-to-grave tracking of the fate and disposal of hazardous wastes from generator to transporter to treatment, storage or disposal. The past disposal of wastes is regulated under the Comprehensive Environmental Response, Compensation and Liability Act (CERCLA), commonly known as Superfund. The Superfund Amendments and Reauthorization Act (SARA) strengthened CERCLA by providing new cleanup standards, requiring cleanup schedules, aiming certain provisions directly at federal facilities, and increasing settlement, liability and enforcement powers for the Environmental Protection Agency (EPA) and private citizens.

The Emergency Planning and Community Right-to-Know Act (SARA Title III) requires facilities using, producing, or storing toxic chemicals to participate in emergency planning committees and to inform the community regarding the safety of these chemicals.

The EPA is responsible for the enforcement of regulations promulgated under RCRA, CERCLA/SARA and Title III of SARA. The Occupational Safety and Health Act (OSH Act) authorizes the Occupational Safety and Health Administration (OSHA), within the Department of Labor (DOL), to set regulations governing the health and safety of workers in the private sector. The Hazardous Materials Transportation Act (HMTA) provides the Department of Transportation (DOT) authority to set regulations governing packaging, handling, labeling, marking, placarding, and routing. EPA has adopted the DOT regulations for hazard communication, packaging and notification, marking, and manifest requirements. The DOT enforces compliance with regulations promulgated under the HMTA. The authority between DOT and OSHA/DOL has not been established in the areas of vehicle operator safety and the protection of workers handling packages containing hazardous material at shipping or transfer facilities. Relationships between the federal authorities, EPA, DOT, and OSHA/DOL are fully described in the final section of this introduction.

The Toxic Substances Control Act (TSCA) provides the EPA with broad authority to regulate chemical substances without regard to specific use. The Clean Air Act (CAA) provides the EPA and authorized states with the authority to set regulations concerning air emissions. The impact of new regulations will require smaller companies to obtain air permits and / or install air pollution control equipment. Note: *where state laws differ from federal ones, the more strenuous regulations may apply.*

RESOURCE CONSERVATION AND RECOVERY ACT

The Resource Conservation and Recovery Act of 1976 (RCRA), as amended by the Hazardous and Solid Waste Amendments of 1984 (HSWA), regulates the multi-faceted problems associated with hazardous waste disposal in this country. The primary objective of the RCRA is to protect human health and the environment. The secondary objective is to conserve valuable material and energy resources by providing assistance to state and local governments for prohibiting open dumping; regulating the management of hazardous wastes; encouraging the recycling, reuse, and treatment of hazardous wastes; and providing guidelines for solid waste management, resource recovery and resource conservation systems. The passage of RCRA closed the loop of environmental protection -- air pollution, water pollution, and now the disposal of hazardous wastes on and in the ground. It provides a cradle-to-grave tracking of the fate and disposal of hazardous wastes from generator to transporter to treatment, storage or disposal. The release or disposal of hazardous substances, including that which took place prior to November 19, 1980 (effective date of the RCRA regulations), is governed by the Comprehensive Environmental Response, Compensation and Liability Act of 1980 (CERCLA, commonly known as Superfund).

Anyone who generates, transports, stores, or disposes of hazardous waste, and anyone who produces, burns, distributes or markets any hazardous waste-derived fuels, or stores hazardous material in underground tanks must comply with RCRA by notifying the U.S. Environmental Protection Agency (EPA) of their activities.

The RCRA statutes, as amended by HSWA, are divided into nine subtitles (A-I). Subtitle C-Hazardous Waste Management (Section 3001-3019) establishes the comprehensive cradle-to-grave program and contains the requirements for RCRA permitting, closure, and post-closure activities. The criteria for the identification of hazardous wastes are also included in Subtitle C. Table 1-1 presents an outline of the EPA regulations codified in Title 40 Code of Federal Regulations (CFR) Parts 260 through 265 and corresponding sections of RCRA.

Table 1-1 CFR/RCRA Correlation

40 CFR	Corresponding RCRA Section and Descriptive Title
Part 260	Definitions generally used in other parts and Provisions generally applicable to other parts
Part 261	Section 3001: Identification and Listing of Hazardous Waste
Part 262	Section 3002: Standards Applicable to Generators of Hazardous Waste
Part 263	Section 3003: Standards Applicable to Transporters of Hazardous Waste
Part 264	Section 3004: Final Standards Applicable to Owners and Operators of Hazardous Waste Treatment, Storage, and Disposal Facilities
Part 265	Section 3004: Interim Status Standards Applicable to Owners and Operators of Hazardous Waste Treatment, Storage, and Disposal Facilities

Part 268 Section 3013: Monitoring, Analysis, and Testing (Land Disposal Restrictions)

Part 270 Section 3005: Permits for Treatment, Storage, or Disposal of Hazardous Waste (EPA Administered Permit Programs: The Hazardous Waste Permit Program)

Part 271 Section 3006: Authorized State Hazardous Waste Program (Requirements for Authorization of State Hazardous Waste Programs)

Part 280 Section 9003: Release Detection, Prevention, and Correction Regulations (Technical Standards and Corrective Action Requirements for Owners and Operators of Underground Storage Tanks (UST))

Part 281 Section 9004: Approval of State Programs (Approval of State Underground Storage Tank Programs)

Citations to EPA regulations are to the Code of Federal Regulations of July 1, 1991, unless otherwise indicated.

Subtitle I to RCRA: Regulation of Underground Storage Tanks

Congress responded in 1984 to the problem of leaking USTs by adding Subtitle I to RCRA. Subtitle I required EPA to develop regulations to protect human health and the environment from leaking USTs. The complete regulations can be found in the Federal Register (September 23, 1988) or by referring to 40 CFR Parts 280 and 281.

Federal-State Relationships under RCRA

RCRA Section 3006 provides authority for states to develop and enforce their own hazardous waste programs in lieu of the EPA-administered federal program. A state must, however, go through an approval process and obtain "state authorization" from EPA to administer its own program.

In order to obtain final authorization the state must demonstrate its own program is "equivalent" to the federal program, that it is "consistent" with federal or state programs in other states, and that it provides adequate capabilities. State programs may be more stringent than the federal program. Thus, knowledge of state laws and regulations is crucial to hazardous waste management.

Almost all states have obtained final authorization to administer RCRA programs. However, only a few states have obtained final authorization to manage mixed hazardous and radioactive wastes.

COMPREHENSIVE ENVIRONMENTAL RESPONSE, COMPENSATION AND LIABILITY ACT

The Comprehensive Environmental Response, Compensation and Liability Act (CERCLA) of 1980 provides a federal mechanism to respond to the hazards posed by any disposal sites and federal authority to respond to current uncontrolled releases of hazardous substances from a vessel (including transportation vehicles) or from any on-shore or off-shore facilities. The act accomplishes this through a scheme of strict liability imposed upon a broad class of potentially responsible parties and the establishment of funding (the "Superfund") which enables the government either to order the responsible parties to undertake a cleanup or to seek reimbursement from the responsible parties when the government has performed the clean-up. Under CERCLA, you may be responsible to clean up your waste (and potentially that of others) even if it is found years later in a contaminated area and even if your own handling was done properly. Thus, it is extremely important to pick a reliable transporter or Treatment Storage or Disposal Facility (TSDF), or have your waste recycled.

CERCLA imposes reporting requirements on owners and operators of vessels and facilities. Releases of reportable quantities of "hazardous substances" must be reported and the responsible party must clean it up. A "hazardous substance" is anything on a "list of lists" compiled by reference to four other major environmental statutes under which toxic or hazardous substances are identified. The Environmental Protection Agency (EPA) is authorized to expand the CERCLA list by adding compounds or mixtures, which, when released into the environment may present substantial danger to public health or welfare or to the environment.

Section 102 of CERCLA required EPA to promulgate regulations establishing Reportable Quantities (RQs) for releases of hazardous substances. For those substances for which EPA has not yet set final reportable quantities, the RQ is one pound. The list of hazardous substances and RQs appears in 40 CFR § 302.4.

Section 105 of CERCLA provides that the government's cleanup activity must be conducted in accordance with the National Contingency Plan (NCP). The NCP establishes a blueprint for cleanup in response to releases to the water, land, and air and assigns response authority to federal and state governments and private parties. The NCP details response procedures, including immediate removal as well as long-term remedial actions. Section 105 also provides the authority for EPA to designate sites for inclusion on the National Priority List (NPL) of sites for the purpose of taking remedial action. Table 1-2 presents an outline of the EPA regulations and corresponding sections of CERCLA.

Table 1-2 CFR\CERCLA Correlation

40 CFR	Corresponding CERCLA Section and Descriptive Title
Part 300	Section 105: National Contingency Plan and National Priorities List
Part 302	Section 102: Designation of Hazardous Substances, Identification of Reportable Quantities and Notification Requirements for Releases

SUPERFUND AMENDMENTS AND REAUTHORIZATION ACT

The Superfund Amendments and Reauthorization Act of 1986 (SARA) strengthened CERCLA by providing new cleanup standards, requiring cleanup schedules, aiming certain provisions directly at federal facilities, and increasing settlement, liability and enforcement powers for the Environmental Protection Agency (EPA) and private citizens.

Section 126 of SARA requires the Secretary of Labor to promulgate regulations for the health, safety, and protection of workers engaged in "hazardous waste operations." Section 126 requires these new standards to include provisions for training, medical surveillance, protective equipment, engineering controls, maximum exposure limits, and decontamination procedures.

SARA Title III

Title III of the Superfund Amendments and Reauthorization Act (SARA), or the Emergency Planning and Community Right-to-Know Act of 1986 was passed in reaction to incidents such as that in Bhopal, India. The Act imposes several different reporting requirements upon facilities using, producing or storing toxic chemicals and mandates the formation of emergency planning entities at the state and local level. Pursuant to Section 302, a new list of toxic chemicals has been created by the EPA, the "extremely hazardous substances" list.

Owners or operators of facilities that have any of the listed extremely hazardous substances present at the facility in excess of the threshold planning quantity established by EPA for that substance must notify the appropriate State Emergency Response Commission (SERC), and the Local Emergency Planning Commission (LEPC), that the facility is subject to Subtitle A of the Act (Emergency Planning and Notification) and must designate a representative to serve as the facility's emergency coordinator. Releases of a reportable quantity of an extremely hazardous substance beyond the boundary of the facility must be reported, as must releases of extremely hazardous substances during transportation.

The Act imposes further reporting requirements in Subtitle B. Owners or operators who are required by the Occupational Safety and Health Administration regulations to prepare or have available Material Safety Data Sheets (MSDS) for hazardous chemicals, must submit copies of all MSDSs or lists of MSDS chemicals to the local emergency planning committee, state emergency response commission, and the local fire department. The owner or operator must also submit annual inventories of MSDS chemicals to the above groups. Finally, owners or operators of facilities in Standard Industrial Codes (SIC) 20-39 must submit annual toxic chemical release forms to EPA on listed toxic chemicals (Section 313 list) manufactured, possessed or used in an amount exceeding the toxic chemical threshold quantity. Table 1-3 presents an outline of the EPA regulations and corresponding sections of SARA Title III.

Subtitle C contains provisions concerning trade secret claims, enforcement, and citizen suits.

TABLE 1.3 SARA Title III Correlation

40 CFR	Corresponding SARA Title III Section and Descriptive Title
Part 355	Section 302, 303, 304, 325, 326, 32 Notification
Part 370	Section 311, 312, 324, 325, 329: Community Right to Know.
Part 372	Section 313: Toxic Chemical Release Reporting Regulations.

OCCUPATIONAL SAFETY AND HEALTH ACT OF 1970

The Occupational Safety and Health Act of 1970 (OSH Act) was passed "to assure so far as possible every working man and woman in the nation safe and healthy working conditions and to preserve our human resources."

Under the act, the Occupational Safety and Health Administration (OSHA) was created within the Department of Labor to:

- encourage employers and employees to reduce workplace hazards and to implement new or improved safety and health programs;
- provide for research in occupational safety and health to develop innovative ways of dealing with occupational safety and health problems;
- establish "separate but dependent responsibilities and rights" for employers and employees to achieve better safety and health conditions;
- maintain a reporting and record-keeping system to monitor job related injuries and illness;
- establish training programs to increase the number and competence of occupational safety and health personnel;

- develop mandatory job safety and health standards and enforce them effectively; and
- provide for the development, analysis, evaluation and approval of state occupational safety and health programs.

In general, coverage of the act extends to all employers and their employees in the 50 states, the District of Columbia, Puerto Rico, and all other territories under United States jurisdiction. Coverage is provided either directly by OSHA or through an OSHA-approved state program.

As defined by the act an employer is any "person engaged in a business affecting commerce who has employees, but does not include the United States or any state or political subdivision of a state." Therefore, the act applies to employers and employees in such varied fields as manufacturing, construction, longshoring, agriculture, law, medicine, charity, disaster relief, organized labor, and private education.

The following are not covered under the act:

- self-employed persons;
- farms at which only immediate members of the farm employer's family are employed; and
- workplaces already protected by other federal agencies under other federal statutes.

But even when another federal agency is authorized to regulate safety and health working conditions in a particular industry, if it does not do so in specific areas, then OSHA standards apply.

OSHA Provisions for Federal Employees

Federal agencies are required under the act to establish and maintain an effective and comprehensive safety and health program. Such a program must be consistent with OSHA standards for private employers. The Secretary of Labor must provide federal agencies with guidance to assist them in maintaining an effective program for their employees. As required by Section 19 of the act and Executive Order 12916, federal agency heads must submit annual reports to the Secretary on the status of their OSHA programs. The Secretary in turn analyzes the reports and statistical data on federal employee injuries and illness and prepares a summary report to the President of the overall findings and recommendations.

Individual agencies may, at their option, establish safety and health committees composed of an equal number of management and employee representatives. Committees have access to agency information on hazards in the workplace and monitor agency performance including inspections. Also, they have the authority to request an OSHA inspection if at least half of the committee is dissatisfied with Agency response to a safety or health problem. OSHA has general inspection authority for agencies which do not set up committees.

OSHA regulations governing federal agency programs appear in 29 CFR Part 1960.

OSHA Jurisdiction/Preemption

Section 4(b)(1) of the OSH Act provides that "[n]othing in this act shall apply to working conditions of employees with respect to which other federal agencies . . . exercise statutory authority to prescribe or enforce standards or regulations affecting occupational safety or health." Therefore, when another federal agency, such as the Department of Transportation (DOT), promulgates safety regulations, the DOT regulations "pre-empt" those of the OSHA. The pre-emption issue is a complex one and is especially difficult with respect to transportation due to recent developments such as the expansion of the OSHA hazard communication standard to the non-manufacturing sector. To date, OSHA and the DOT have not been able to complete a Memorandum of Understanding regarding the jurisdiction of each agency over the health and safety of transportation workers.

HAZARDOUS MATERIALS TRANSPORTATION ACT

The Hazardous Materials Transportation Act (HMTA) was enacted in 1975 in response to public concerns over the fragmentation of the then existing regulatory structure and the lack of enforcement. The HMTA gave the Secretary of Transportation broad authority to set regulations applicable to all modes of transportation. The HMTA expanded the Department of Transportation's (DOT's) potential jurisdiction to any traffic "affecting interstate commerce;" authorized the Secretary to designate materials as hazardous; permitted the Secretary to issue regulations governing packaging, handling, labeling, marking, placarding and routing; authorized DOT to grant exemptions; and authorized DOT to assess criminal and civil penalties for noncompliance. The HMTA provides for federal preemption of non-federal rules found to be "inconsistent" with the federal regulations.

Overview of the Hazardous Materials Transportation Regulations (HMTR)

The regulations of DOT which pertain to the transportation of hazardous materials fall under Subtitle B Chapter 1 , Subchapter C of Title 49 CFR. Table 1-4 summarizes the contents of the regulations with reference to Parts of 49 CFR.

Part 171 of the Hazardous Materials Transportation Regulations (HMTR) outlines general information relating to the transportation requirements of hazardous materials. This part includes definitions (§ 171.8), import and export shipments (§ 171.12), and hazardous materials incidents (§§ 171.15 and 171.16) reporting requirements.

Part 173 provides the requirements for the preparation of hazardous materials for shipment. It designates the standards for packaging, quantity limitation, and the use and re-use of hazardous materials shipping containers.

Part 172 includes the Hazardous Materials Table (HMT). The HMT lists those materials which have been designated for transportation purposes as hazardous materials. The HMT also identifies the proper shipping names, hazard class, and labeling, packaging and shipping requirements (49 CFR § 172.101). The HMT includes hazardous wastes.

Part 172 also contains extensive information on the communication regulations regarding hazardous materials. Requirements for shipping papers (49 CFR §§ 172.200 - 172.205), marking (§§ 172.300 - 172.338), labeling (§§ 172.400 - 172.450), and placarding (§§ 172.500-172.558) are included in this part.

A key component of the packaging requirements in Part 173 sets forth the hazard class hierarchy used in properly classifying materials having more than one hazard.

TABLE 1-4 Summary of U. S. DOT Hazardous Materials Regulations

49 CFR	Topical heading
Part 171	General information, regulations and definitions
Part 172	Hazardous materials table and hazardous materials communications regulations
Part 173	Shippers - General requirements for shipments and packaging
Part 174	Carriage by rail
Part 175	Carriage by aircraft
Part 176	Carriage by vessel
Part 177	Carriage by public highway
Part 178	Shipping container specifications
Part 179	Specifications for tank cars

Operating, handling, and loading requirements for hazardous materials shipped by various modes of transport are included in Parts 174 through 177 of the HMTR. Rail transport is governed by Part 174, aircraft by Part 175, and vessel by Part 176.

Part 177 of the HMTR governs the transportation of hazardous material by public highway. In addition to the general, loading and unloading requirements, there is an extensive chart of loading and storage information (§ 177.848). This chart delineates which hazardous materials must not be loaded, stored or transported together. Part 177 also sets standards by which hazardous materials may be carried on motor vehicles which transport passengers for hire and the requirements for motor vehicle operators involved in an accident while transporting hazardous materials.

Parts 178 and 179 contain detailed requirements setting specifications for shipping containers (Part 178) and tank cars (Part 179). These specifications include standards for the quality of materials used for construction, methods of construction, sealing and testing. These two parts of the HMTR are extensive and contain over 500 pages of specifications.

TOXIC SUBSTANCES CONTROL ACT

The 1976 Toxic Substances Control Act (TSCA) was designed as a catchall to close the loopholes in the environmental protection and chemical manufacture use laws. It gives the United States Environmental Protection Agency (EPA) broad authority to regulate chemical substances without regard to specific use (e.g., food, drug, cosmetic) or area of application (e.g., food crops) if they present a hazard to health or the environment. The law controls the chemical at its source before it is dispersed into the environment (where environmental protection laws are employed). Excluded from coverage under TSCA are food, food additives, drugs, or cosmetics and nuclear materials regulated by the Atomic Energy Act.

Asbestos is one of the key contaminants controlled under TSCA. The adverse health effects associated with asbestos exposure has been extensively studied for many years. Exposure of asbestos fibers into the respiratory system can result in asbestosis (scarring and reduced lung capacity) and mesothelioma (lung cancer).

The Asbestos Hazard Emergency Response Act (AHERA) was adopted under TSCA to establish regulations for training, inspection, handling, and disposal of asbestos. While developed to address asbestos hazards in schools, the regulations are often applied to public and publicly-accessible privately-owned facilities.

Originally (prior to 1976) chlorofluorocarbons (CFCs) were not classified as air pollutants and did not pose a recognized hazard in the workplace; there was no means of regulating their use. Subsequently they have been regulated as ozone-depleting chemicals by the Clean Air Act. Until passage of the Clean Air Act that included regulation of CFCs, TSCA was used by the EPA to control environmentally sensitive substances such as CFCs at the point of manufacture. TSCA is still used to control environmentally sensitive substances which are not regulated by other acts.

Polychlorinated biphenyls (PCBs) were commonly used in many commercial operations including transformers, hydraulic fluids, and lubricants. Human exposure to PCB (and associated chemicals such as dioxins and furans) has been determined to pose high cancer risks. To address these risks, TSCA incorporated regulations that ban the manufacture of PCBs.

TSCA is unique in that it is designed as a gap-filling law. The EPA defers to other agencies for action if those agencies having statutory authority under another law are dealing with identified problems. When the EPA has sufficient authority under another law (e.g., CAA, CWA, RCRA, etc.), the EPA is directed to use the other law rather than the gap-filling TSCA.

CLEAN AIR ACT

On November 15, 1990, President Bush signed into law sweeping revisions of the Clean Air Act (CAA) which was originally signed into law in 1970. The new amendments contain titles that:

- strengthen measures for attaining air quality standards (Title I),
- establish tighter emission standards for vehicles and fuels (Title II),
- expand the regulation of hazardous air pollutants (Title III),
- require substantial reductions in power plant emissions for control of acid rain (Title IV),
- establish operating permits for all major sources of air pollution (Title V),
- establish provisions for stratospheric ozone protection (Title VI), and
- expand enforcement powers and penalties (Title VII).

The CAA amendments will have far-reaching effects not only on environmental compliance activities at industrial facilities, but also on procurement, maintenance, and motor vehicle operation activities. Some of the more important provisions of the 1990 CAA amendments are discussed below.

Nonattainment Provisions

The original 1970 CAA authorized the EPA to establish National Ambient Air Quality Standards (NAAQS) to limit levels of pollutants in the air. To date EPA has promulgated NAAQS for sulfur dioxide, nitrogen dioxide, carbon monoxide, ozone, lead, and particulate matter (PM-10); collectively called criteria pollutants. All areas of the United States are required to maintain ambient levels of these pollutants below the ceilings established by the NAAQS; any area that does not meet these standards is a "nonattainment" area (NAA).

Under previous law, the geographic units for which states were required to submit NAA data were not specified. The Amendments, however, require that the boundaries of serious, severe, or extreme ozone or carbon monoxide nonattainment areas located within Metropolitan

Statistical Areas (MSAs) or Consolidated Metropolitan Statistical Areas (CMSAs), be expanded to include the entire MSA or CMSA. This will lead to an expansion of geographic areas classified as nonattainment, because all urban counties included in an affected MSA of CMSA, regardless of their attainment status, will now become part of the NAA.

The new definitions of "major" sources are a significant aspect of the nonattainment classification scheme. Under previous law, "major" sources were those with the potential to emit over 100 tons per year (TPY) with emission controls of a criteria pollutant. Existing major sources in specified industrial categories in NAAs could be required to retrofit by installing "reasonably available control technology." New major sources had to install very stringent "lowest achievable emission rate" technology and acquire 1:1 offsets to obtain construction permits. The CAA Amendments reduce the size of plants subject to permitting and stringent retrofitting or offsetting requirements as follows:

- In serious ozone NAAs, major sources are now those with the potential to emit over 50 TPY of volatile organic compounds.
- In severe ozone NAAs, major sources are now those that emit 25 TPY and in extreme areas, 10 TPY.
- For serious carbon monoxide NAAs, a major source is now one that emits 50 TPY.
- For serious PM-10 NAAs, a major source is now one that emits 70 TPY.

Hazardous Air Pollutants

Congress expanded the number of hazardous air pollutants (HAPs) from eight to 189, and the EPA Administrator must periodically review and revise the list and add other pollutants that present a major threat of adverse human health or environmental effects. In addition, the Amendments require a major shift in approach from regulation of HAPs using health-based, substance-specific standards, to regulation under technology-based standards applicable to categories of emission source rather than to the substances emitted.

More sources of HAPs, including small sources, will come under operating permit requirements because of the new definitions of "major source" and "area source." For the regulation of HAPs, a "major source" is any stationary source or group of sources located in a contiguous area and under common control that emits or has the potential to emit (with emission controls), in the aggregate, 10 TPY or more of any single HAP or 25 TPY or more of any combination of HAPs. The EPA Administrator may establish lesser quantities or, for radionuclides, different criteria for a major source. "Area sources" -- any stationary sources of HAPs that are not major sources (excluding motor vehicles and non-road vehicles) -- may also come under regulation in the future.

FEDERAL AGENCY INTERACTIONS OVERVIEW

This section describes briefly the coordination of regulatory activities between the Department of Transportation (DOT), the Environmental Protection Agency (EPA), and Occupational Safety and Health Administration (OSHA).

DOT/EPA Interactions

Section 3003 of the Resource Conservation and Recovery Act (RCRA) directed EPA to establish standards for transporters of hazardous waste and to coordinate regulatory activities with the DOT. Pursuant to this statutory directive, EPA adopted the DOT regulations for hazard communication, packaging, and reporting discharges, and has enacted additional notification, marking, manifest and cleanup requirements. The relationship between RCRA regulations and Hazardous Material Transportation Regulations (HMTR) is summarized in Table 1-5.

TABLE 1-5 Cross-Reference of EPA and DOT Regulations

Topics	40 CFR Regulations	49 CFR Regulations
Definitions	Section 260.10	Section 171.8
Identification and listing of hazardous materials and wastes	Part 261	Section 172. 1
Characteristics of hazardous materials and wastes	Sections 261.20 to 261.24	Section 171.8, Part 173
Compliance with manifest system	Sections 262.20 to 262.23 and Sections 263.20 to 263.21	Section 172.205
Packaging and containers	Section 262.30	Parts 173,178, and 179
Labeling requirements	Section 262.31	Section 172.400
Marking requirements	Section 262.32	Section 172.300 to 172.330
Placarding requirements	Section 262.33	Sections 172.500 to 172.558
Hazardous material and waste discharge incidents	Sections 263.30 and 263.31	Sections 171.15 to 171.16

The EPA and DOT have entered into a Memorandum of Understanding (MOU) [45 Fed. Reg. 51645 (Aug. 4, 1980)] that delineates the enforcement responsibility of EPA and DOT. The MOU authorizes EPA to:

- conduct an ongoing program to monitor the compliance of generators of hazardous waste and hazardous waste management facilities with the RCRA regulations;
- bring enforcement actions involving hazardous waste transporters where the transportation is ancillary to treatment, storage or disposal of hazardous waste or other activities normally under the jurisdiction of the EPA. For example, a "midnight dumper" will be considered an illegal disposer. The fact that the "dumper" is transporting the waste is ancillary to the disposal of the waste and EPA will bring appropriate enforcement action against him;
- provide DOT with current lists of all hazardous waste transporters who have notified EPA as required in Section 3010 of RCRA;
- investigate reports from DOT which give EPA cause to suspect that a violation of RCRA has occurred and where warranted, initiate appropriate regulatory or enforcement action under RCRA;
- bring enforcement actions to address hazardous waste activities which may present an "imminent and substantial endangerment to health and the environment" as those words are used and administered by EPA (Section 7003 of RCRA and Section 504 of the Clean Water Act).

The MOU also authorizes the DOT to:

- conduct an ongoing program of inspections of transporters and shippers of hazardous waste to monitor their compliance with Hazardous Materials Transportation Regulations (HMTR);
- investigate reports from EPA which give DOT cause to suspect that a violation of the Hazardous Materials Transportation Act (HMTA) has occurred and, where warranted, initiate appropriate regulatory or enforcement action under HMTA.

In addition to outlining the foregoing responsibilities, the MOU requires each agency to report actions, coordinate investigations and enforcement, exchange pertinent information and reports, maintain a close working relationship, and assign liaison representatives to offices at the regional and central headquarter levels.

DOT/OSHA Interactions

The Occupational Safety and Health (OSH Act) prohibits the Occupational Safety and Health Administration (OSHA) from exercising regulatory authority over working conditions of employees where another federal agency has already exercised its regulatory authority. Two major regulatory areas of concern arise with respect to transportation-vehicle operator safety and the protection of workers handling packages containing hazardous materials at shipping or transfer facilities. OSHA has not taken any regulatory action with respect to vehicle operators because DOT has established requirements. OSHA has generally accepted DOT's packaging rules.

The OSHA hazard communication standard requires labeling of containers that are used in or leave the workplace in a manner that does not conflict with DOT regulations. The hazard communication standard has been expanded to include all employers with employees routinely exposed to hazardous chemicals in their workplace [52 Fed. Reg. 31852 (Aug. 24, 1987)].

SUMMARY

The following acts impact the regulation of hazardous and/or toxic waste: RCRA, CERCLA (SARA), OSHA, TSCA and HMTA. Additional acts may affect product substitution (e.g., Clean Air Act) and/or discharges from processes (Clean Water Act). EPA, OSHA, and DOT have the authority to enforce these laws. The laws or statutes are put into action by regulations appearing in the Federal Register (Fed. Reg.) and subsequently codified into the Code of Federal Regulations (CFR). Additionally, each state has its own hazardous waste regulations which must be as rigorous as federal regulations and may be more restrictive than federal regulations.

CHAPTER 2

RCRA FACILITY ADMINISTRATOR'S RESPONSIBILITIES

Every business or institution which generates hazardous waste must have a person responsible for the administration of a hazardous waste management program. In this book, that individual will be referred to as the RCRA administrator.

In many cases, the RCRA administrator will not only be responsible for hazardous waste management but will also coordinate the business or institution's overall environmental compliance.

This book will only deal with matters directly applicable to *hazardous waste* management and compliance, but the guidance given for the selection of a RCRA administrator is pertinent to the selection of an overall environmental compliance manager.

SELECTION OF AN RCRA ADMINISTRATOR

For a larger company or government facility, it is of paramount importance to select a competent RCRA administrator and to vest that individual with sufficient power and access to resources to carry out his (or her) responsibilities.

A RCRA program administrator should meet the following basic requirements:

- He or she should have a technical background, preferably one of the basic sciences, especially chemistry;
- He or she should be able to make decisions, lead and delegate responsibility;
- In addition, the employer would be wise to select a candidate who has some direct or indirect knowledge and experience with the processes that generate the majority of the facility's waste; and
- The program administrator candidate should be thorough and adept at interfacing with inter-company or facility management and government regulators.

The RCRA Program Administrator should have access to upper level management. The administrator should have the clear backing of upper level management. Upper level management must have a clear policy stating their environmental compliance objectives that the administrator and generator employees can understand. Upper level management must communicate to line managers through policy or directive to cooperate with the program initiated by the administrator. Finally, the administrator must be provided with a reasonable budget. The job will be difficult enough with upper level management's visible support; it will be nearly impossible without it.

A company should expect to pay its RCRA administrator a reasonable percentage more than comparable technical managers. The administrator will be assuming significant liability. Qualified administrators are in demand and it is in the company's best interest to attract and maintain a competent administrator.

The administrator should be expected to retain a staff, which at the minimum consists of a secretary / administrative assistant and a field supervisor or industrial waste specialist, who can effectively interact with generator employees and has a strong back.

RESPONSIBILITIES

The RCRA administrator should initiate the following actions and programs in stated order, unless they already exist, in which case he should review the program for efficiency, thoroughness and regulatory compliance. The bulk of this book will focus on how to complete these tasks.

1. The administrator must determine the facility's needs with respect to hazardous waste management. To do this he must first conduct an initial waste survey to determine the relative magnitude of the waste being generated or likely to be generated, the types of waste generated and the state of regulatory compliance.

2. The administrator must determine the facility's likely storage needs after the initial waste survey. He should be cognizant of existing satellite accumulation areas, the need for additional areas, 90-day accumulation areas and the need for a permanent storage facility.

3. If the facility is regulated (i.e., not a conditionally exempt small quantity generator) both for federal and state purposes, the administrator should contact the EPA and/or the appropriate state agencies to obtain necessary reporting and permitting forms.

4. The administrator must complete a Notification of Hazardous Waste Activity form (EPA Form 8700-12) to obtain an EPA establishment ID number for each geographically distinct facility owned by the company or managed by the entity.

5. The administrator must establish a unified generator waste profiling system for hazardous and some non-hazardous waste. The waste profiles are the cornerstone of the waste tracking, reporting, treatment, disposal and regulatory compliance programs. Each generator or his environmental compliance officer, designee or consultant must complete a profile for each hazardous waste stream, potential hazardous waste stream or large quantity non-hazardous waste stream.

6. A paper-based and electronic database recordkeeping system should be established. The records kept should include:

   - generator waste stream profiles
   - all manifests
   - TSDs used
   - current storage facility inventory
   - all exception reports
   - biennial reports
   - SARA Title III reports

   It cannot be overly stressed that accurate recordkeeping will avoid potential regulatory non-compliance and reduce liabilities. All large quantity generators that have 10 or more generators or 10 or more waste streams should develop an electronic database system.

7. All waste profiles should be submitted to the administrator. They must be reviewed for completeness, need for additional analytical analysis and to determine whether the waste

stream is a RCRA hazardous waste, a DOT hazardous substance, a TSCA waste or a non-hazardous waste stream.

8. The administrator must establish and conduct or contract for generator training for all generator employees actively participating in the generation, storage or transportation of hazardous waste. The training should support or be an adjunct to waste profiling, siting of waste accumulation areas, storage and shipment of waste. If the facility has a permitted on-site storage facility, employees of that facility will require OSHA training.

9. The administrator must contact and evaluate potential shipping, treatment, storage and disposal companies to handle waste. Each candidate should be carefully selected.

10. Internal audits of generator practices, storage areas and recordkeeping should be conducted at least annually.

11. The administrator should evaluate his program needs and available time and resources to determine if outside consultants are needed.

12. The administrator should perform a waste minimization assessment at least annually.

13. The administrator should make periodic reports to management or prepare resources needed for compliance in the short and long term.

CHAPTER 3

INITIAL DETERMINATION OF HAZARDOUS WASTE STORAGE, TREATMENT, AND DISPOSAL NEEDS

In this discussion of the initial determination of a facility's hazardous waste, storage, treatment and disposal needs, the assumption is that the facility has not applied for a Part A Permit (notification by completion of EPA form 8700-72) or Part B Permit (for permitted treatment, storage, and/or disposal facilities). It is unlikely that a facility has not applied for one or both permits, but for the sake of presenting a step-by-step plan we assume that the RCRA administrator has to complete all steps of the process.

The review of the process will be valuable for those individuals who will audit the program or will in a later iteration produce a clearly documented paper trail for the determination of the facility's regulatory status.

GATHERING INFORMATION

The administrator should obtain an aerial photograph, if possible, and a map showing all the buildings, treatment units, and generation units located in the geographic facility. If there are separate geographic facilities, he should obtain the above photographs and maps for each. A separate geographic facility is one that is not continuous with another, or is separated by a public highway. A copy of the map should be to such a scale that notations can be written on it listing the building addresses, major operations and hazardous waste generating groups.

The administrator designee should obtain the above information through corporate records, knowledgeable individuals or by physical inspections. Once the above information has been obtained, each generating group in each separate facility should be contacted and the responsible party asked about the nature of the operation generating the waste, and for a ball-park idea of the quantity of waste (in pounds or kilograms for solids and gallons or liters for liquids) that the generating unit produces per month.

It may be difficult at this juncture to decide if the waste generated is hazardous or not. If the administrator suspects that the waste may be hazardous, he should initially count it as hazardous waste.

The administrator will need to obtain a separate EPA Establishment I.D. Number for each geographically distinct generating facility. If there are questions concerning the status of a facility (e.g., whether it is geographically distinct or not), the questions should be addressed to the appropriate regulatory agency.

DETERMINING THE FACILITY'S STATUS

After the survey, the administrator should have the necessary information to decide the status of main geographic generators and geographically distinct facilities. There are three categories: large quantity generators, small quantity generators and conditionally exempt small quantity generator facilities.

A large quantity generator will produce more than 1,000 kilograms of hazardous waste per month or more than 1 kilogram of acutely hazardous waste per month.

A small quantity generator will produce less than 1,000 kilograms but more than 100 kilograms per month and no more than 1 kilogram of acutely hazardous waste per month. A small quantity generator can accumulate no more than 6000 kilograms of hazardous waste in any 180-day period (270 days if more than 200 miles to a licensed hazardous waste facility) without a special storage permit.

A conditionally exempt small quantity generator will produce less than 100 kilograms per month and less than 1 kilogram of acutely hazardous waste and not accumulate more than 1,000 kilograms of hazardous waste at any time.

These are federal categories. The particular state that the generating facility is located in may have more stringent requirements.

The information obtained from the initial survey will provide the basis for determining generator categories, the approximate storage needs, and if current treatment capabilities were assessed, the approximate disposal needs and most importantly, whether a permitted storage facility is needed. If a permitted storage facility is needed, that will necessitate the completion of Part A and B permit applications.

CHAPTER 4

NOTIFICATION OF REGULATORY AGENCIES (PART A PERMITS)

The initial vehicle for notifying regulatory authorities concerning hazardous waste activity is EPA Form 8700-12. Consequently we will present that form and address issues surrounding the completion and submittal of the form. It is important to understand that EPA Form 8700-12 addresses the requirements of the Federal hazardous waste program. Many states which have EPA authorization may have requirements that are different and more stringent than the Federal requirements. It is up to the generator, transporter or treatment, storage and disposal facility to determine State and Federal requirements.

WHO SHOULD COMPLETE EPA FORM 8700-12?

1. A potential hazardous waste generator;
2. Hazardous waste transporters;
3. Owner or operator of a facility that treats, stores, or disposes of hazardous waste (must also complete Part B, see chapter 4);
4. A potential marketer or burner of hazardous waste or used oil burned for purposes of energy recovery.

After receipt of a completed 8700-12 the EPA will issue a U.S. EPA Identification Number.

How do you decide if you handle a regulated hazardous waste? Refer to Chapter 8 of this book or to 40 CFR Part 261 of the Code of Federal Regulations to decide if you have regulated hazardous waste, then contact the appropriate state agency for state requirements. A list of state agencies is included in this chapter.

A note for persons who market or burn hazardous waste or used oil and any material produced from or otherwise containing hazardous waste or used oil for energy recovery: You are required to notify the U.S. EPA, or the state agency if the state is authorized to operate its own hazardous waste program, and obtain a U.S. EPA Identification Number unless your materials are exempt from regulations. Hazardous waste and used oil are considered to be burned for energy recovery if they are burned in a boiler or industrial furnace that is not regulated as a hazardous waste incinerator. If you have notified the U.S. EPA of hazardous waste activities, you must re-notify the agency if you have waste-as-fuel activities. Your EPA ID number will not change.

WHO IS EXEMPT FROM WASTE-AS-FUEL NOTIFICATION REQUIREMENTS

1. Ordinary generators (and initial transporters): Generators (and initial transporters who pick up used oil or hazardous waste from generators) are not marketers subject to the notification requirement if they do not market hazardous waste fuel or used oil fuel directly to a burner. In such situations, it is the recipient of that fuel who makes the decision to market the materials as a fuel (typically after processing or blending), and it is the recipient who must notify the EPA.

   In addition, used oil generators or initial transporters who send their oil to a person who processes or blends it to produce used oil fuel and who incidentally burns used oil to provide energy for the processing or blending are also exempt from the notification requirement. This is because such persons are generally considered to be primarily fuel processors and marketers, but only incidental burners.

2. Persons Who Market or Burn Specification Used Oil Fuel: Used oil fuel that meets the specification provided under 40 CFR 266.40(e) is essentially exempt from the regulations.

However, the person who first claims that the used oil meets the specification is subject to the notification and certain other requirements. The burner (or any subsequent marketer) is not required to notify the EPA.

3. Used Oil Generators Operating Used-Oil-Fired Space Heaters: Persons who burn their used oil (and used oil received from individuals who are do-it-yourself oil changers) in used-oil-fired space heaters are exempt from the notification requirement provided that the device is vented to the outdoors.

4. Specific Exemptions Provided by 40 CFR 261.6: The rules provide conditional exemptions for several specific waste-derived fuels under 261.6(a)(3), including fuels produced by petroleum refineries that recycle refinery hazardous waste, and coke and coal tar derived from coal coking wastes by the iron and steel industry. Marketers and burners of these exempted fuels are not subject to the notification requirement. (Refer to used oil chapter.)

A person who is subject to the hazardous waste regulations should submit one notification (form 8700-12) per site or location.

When the owner of a facility changes, the new owner must notify the U.S. EPA of the change, even if the previous owner already received a U.S. EPA Identification Number. Because the EPA I.D. number is "site-specific" the new owner will keep the existing I.D. number. If the facility moves to another location, the owner/operator must notify the EPA of this change. In that instance a new U.S. EPA I.D. number will be assigned since the facility has changed locations.

ADDITIONAL POINTS

1. Contact your state hazardous waste management agency to see if they require a different document than the federal form included here. State forms will almost always require the same data as the federal form, but may have additional requirements.

2. Complete one copy of the form for *each* of your plant sites or business locations where hazardous waste is generated. Each site or location should obtain an EPA identification number.

3. Send the form to your *state* agency; they will forward a copy to the EPA. You will soon receive in the mail your U.S. EPA ID number for use on all hazardous waste manifests originating at the site.

4. If you move the location of your business, contact your state environmental agency. If there was a previous business at your new site using an EPA ID number, you will be assigned that number. In other words, the number stays with the address. If no previous owner at the site was assigned an EPA ID number, you will receive a new number for your new location.

Please refer to the *Instructions for Filing Notification* before completing this form. The information requested here is required by law *(Section 3010 of the Resource Conservation and Recovery Act).*

EPA Notification of Regulated Waste Activity

United States Environmental Protection Agency

Date Received (For Official Use Only)

I. Installation's EPA ID Number *(Mark 'X' in the appropriate box)*

A. First Notification ☐

B. Subsequent Notification *(complete item C)* ☐

C. Installation's EPA ID Number

II. Name of Installation *(Include company and specific site name)*

III. Location of Installation *(Physical address not P.O. Box or Route Number)*

Street

Street (continued)

City or Town | State | ZIP Code

County Code | County Name

IV. Installation Mailing Address *(See instructions)*

Street or P.O. Box

City or Town | State | ZIP Code

V. Installation Contact *(Person to be contacted regarding waste activities at site)*

Name *(last)* | *(first)*

Job Title | Phone Number *(area code and number)*

VI. Installation Contact Address *(See instructions)*

A. Contact Address: Location ☐ Mailing ☐

B. Street or P.O. Box

City or Town | State | ZIP Code

VII. Ownership *(See instructions)*

A. Name of Installation's Legal Owner

Street, P.O. Box, or Route Number

City or Town | State | ZIP Code

Phone Number *(area code and number)* | B. Land Type | C. Owner Type | D. Change of Owner Indicator: Yes ☐ No ☐ | (Date Changed) Month Day Year

Continue on rever

ID – For Official Use Only

VIII. Type of Regulated Waste Activity *(Mark 'X' in the appropriate boxes. Refer to instructions.)*

A. Hazardous Waste Activity

1. Generator (See instructions)
   - ☐ a. Greater than 1000kg/mo (2,200 lbs.)
   - ☐ b. 100 to 1000 kg/mo (220 – 2,200 lbs.)
   - ☐ c. Less than 100 kg/mo (220 lbs.)
2. Transporter (Indicate Mode in boxes 1-5 below)
   - ☐ a. For own waste only
   - ☐ b. For commercial purposes

   Mode of Transportation
   - ☐ 1. Air
   - ☐ 2. Rail
   - ☐ 3. Highway
   - ☐ 4. Water
   - ☐ 5. Other – specify

☐ 3. Treater, Storer, Disposer (at installation) Note: A permit is required for this activity; see instructions.

4. Hazardous Waste Fuel
   - ☐ a. Generator Marketing to Burner
   - ☐ b. Other Marketers
   - ☐ c. Boiler and/or Industrial Furnace
     - ☐ 1. Smelter Deferral
     - ☐ 2. Small Quantity Exemption

     Indicate Type of Combustion Device(s)
     - ☐ 1. Utility Boiler
     - ☐ 2. Industrial Boiler
     - ☐ 3. Industrial Furnace

☐ 5. Underground Injection Control

B. Used Oil Fuel Activities

1. Off-Specification Used Oil Fuel
   - ☐ a. Generator Marketing to Burner
   - ☐ b. Other Marketer
   - ☐ c. Burner – indicate device(s) – Type of Combustion Device
     - ☐ 1. Utility Boiler
     - ☐ 2. Industrial Boiler
     - ☐ 3. Industrial Furnace

☐ 2. Specification Used Oil Fuel Marketer (or On-site Burner) Who First Claims the Oil Meets the Specification

IX. Description of Regulated Wastes *(Use additional sheets if necessary)*

A. Characteristics of Nonlisted Hazardous Wastes. Mark 'X' in the boxes corresponding to the characteristics of nonlisted hazardous wastes your installation handles. *(See 40 CFR Parts 261.20 – 261.24)*

1. Ignitable *(D001)*	2. Corrosive *(D002)*	3. Reactive *(D003)*	4. Toxicity Characteristic *(D000)*	(List specific EPA hazardous waste number(s) for the Toxicity characteristic contaminant(s))
☐	☐	☐	☐	

B. Listed Hazardous Wastes. (See 40 CFR 261.31 – 33. See instructions if you need to list more than 12 waste codes.)

1	2	3	4	5	6
7	**8**	**9**	**10**	**11**	**12**

C. Other Wastes. (State or other wastes requiring a handler to have an I.D. number. See instructions.)

1	2	3	4	5	6

X. Certification

I certify under penalty of law that this document and all attachments were prepared under my direction or supervision in accordance with a system designed to assure that qualified personnel properly gather and evaluate the information submitted. Based on my inquiry of the person or persons who manage the system, or those persons directly responsible for gathering the information, the information submitted is, to the best of my knowledge and belief, true, accurate, and complete. I am aware that there are significant penalties for submitting false information, including the possibility of fine and imprisonment for knowing violations.

Signature	Name and Official Title *(type or print)*	Date Signed

XI. Comments

Note: Mail completed form to the appropriate EPA Regional or State Office. (See Section III of the booklet for addresses.)

ID - For Official Use Only

IX. Description of Regulated Wastes Continued *(Additional sheet)*

B. Listed Hazardous Wastes. (See 40 CFR 261.31 - 33. Use this page only if you need to list more than 12 waste codes.)

13	14	15	16	17	18
19	20	21	22	23	24
25	26	27	28	29	30
31	32	33	34	35	36
37	38	39	40	41	42
43	44	45	46	47	48
49	50	51	52	53	54
55	56	57	58	59	60
61	62	63	64	65	66
67	68	69	70	71	72
73	74	75	76	77	78
79	80	81	82	83	84
85	86	87	88	89	90
91	92	93	94	95	96
97	98	99	100	101	102
103	104	105	106	107	108
109	110	111	112	113	114
115	116	117	118	119	120

LINE-BY-LINE INSTRUCTIONS FOR COMPLETING EPA FORM 8700-12

Type or print in black ink all items except Item X, "Signature," leaving a blank box between words. The boxes are spaced at 1/4" intervals which accommodate elite type (12 characters per inch). When typing, hit the space bar twice between characters. If you print, place each character in a box. Abbreviate if necessary to stay within the number of boxes allowed for each item. If you must use additional sheets, indicate clearly the number of the item on the form to which the information on the separate sheet applies.

(Note: When submitting a subsequent notification form, notifiers must complete in their entirety Items I, II, III, VI, VIII and X. Other sections that are being added to (i.e., newly regulated activities) or altered (i.e., installation contact) must also be completed. All other sections may be left blank.)

Item I -- Installation's EPA ID Number

Place an "X" in the appropriate box to indicate whether this is your first or a subsequent notification *for this site*. If you have filed a previous notification, enter the EPA Identification Number assigned to this site in the boxes provided. Leave EPA ID Number blank if this is your first notification *for this site*.

Note: When the owner of a facility changes, the new owner must notify U.S. EPA of the change, even if the previous owner already received a U.S. EPA Identification Number. Because the U.S. EPA ID Number is "site-specific," the new owner will keep the existing ID number. If the business moves to another location, the owner/operator must notify EPA of this change. In this instance, a new U.S. EPA Identification Number will be assigned, since the facility has changed locations.

Items II and III -- Name and Location of Installation

Complete Items II and III. Please note that the address you give for Item III, "Location of Installation," must be a physical address, *not a post office box or route number*.

County Name and Code: Give the county code, if known. If you do not know the county code, enter the county name, from which EPA can automatically generate the county code. If the county name is unknown, contact your local Post Office. To obtain a list of county codes, contact the National Technical Information Service, U.S. Department of Commerce, Springfield, Virginia, 22161 or call (703) 487-4650. The list of codes is contained in the Federal Information Processing Standards Publication (FIPS PUB) number 6-3.

Item IV -- Installation Mailing Address

Please enter the Installation Mailing Address. If the Mailing Address and the Location of Installation (Item III) are the same, you can print "Same" in the box for Item IV.

Item V -- Installation Contact

Enter the name, title, and business telephone number of the person who should be contacted regarding information submitted on this form.

Item VI -- Installation Contact Address

A) Code: If the contact address is the same as the location of installation address listed in Item III or the installation mailing address listed in Item IV, place an "X" in the appropriate box to indicate where the contact may be reached.

B) Address: Enter the contact address *only* if the contact address is different from either the location of installation address (Item III) or the installation mailing address (Item IV), and Item VI. A. was left blank.

Item VII -- Ownership

A) Name: Enter the name of the legal owner(s) of the installation, including the property owner. Also enter the address and phone number where this individual can be reached. Use the comment section in XI or additional sheets if necessary to list more than one owner.

B) Land Type: Using the codes listed below, indicate in VII.B. the code which best describes the current legal status of the land on which the facility is located:

F = Federal
S = State
I = Indian
P = Private
C = County
M = Municipal*
D = District
O = Other

**Note: If the Land Type is best described as Indian, County or District, please use those codes. Otherwise, use Municipal.*

C) Owner Type: Again, using the codes listed above in VII.B., indicate in VII.C. the code which *best describes* the legal status of the current owner of the facility:

D) Change of Owner Indicator: *(If this is your installation's first notification, leave Item VII.D. blank and skip to Item VIII. If this is a subsequent notification, complete Item VII.D. as directed below)*

If the owner of this facility has changed since the facility's original notification, place an "X" in the box marked "Yes" and enter the date the owner changed.

If the owner of this facility has not changed since the facility's original notification, place an "X" in the box marked "No" and skip to Item VIII.

If an additional owner(s) has been added or replaced since the facility's original notification, place an "X" in the box marked "Yes". Use the comment section in XI to list any additional owners, the dates they became owners, and which owner(s) (if any) they replaced. If necessary attach a separate sheet of paper.

Item VIII -- Type of Regulated Waste Activity

A) Hazardous Waste Activity: Mark an "X" in the appropriate box(es) to show which hazardous waste activities are going on *at this installation*.

1. Generator: If you generate a hazardous waste that is identified by characteristic or listed in 40 CFR Part 261, mark an "X" in the appropriate box for the quantity of non-acutely hazardous waste that is generated per calendar month. If you generate acutely hazardous waste please refer to 40 CFR Part 262 for further information.

2. Transporter: If you transport hazardous waste, indicate if it is your own waste, for commercial purposes, or mark both boxes if both classifications apply. Mark an "X" in each appropriate box to indicate the method(s) of transportation you use. Transporters do not have to complete Item IX of this form, but must sign the certification in Item X. The federal regulations for hazardous waste transporters are found in 40 CFR, Part 263.

3. Treater/Storer/Disposer: If you treat, store or dispose of hazardous waste, then mark an "X" in this box. You are reminded to contact the appropriate addressee listed for your state in Section III.C. to request Part A of the RCRA Permit Application. The federal regulations for hazardous waste facility owners/operators are found in 40 CFR Parts 264 and 265.

4. Hazardous Waste Fuel: If you market hazardous waste fuel, place an "X" in the appropriate box(es). If you burn hazardous waste fuel on-site, place an "X" in the appropriate box and indicate the type(s) of combustion devices in which hazardous waste fuel is burned. (Refer to definition section for complete description of each device.)

"Other Marketer" is defined as any person, other than a generator marketing hazardous waste, who markets hazardous waste fuel.

Note: Generators are required to notify for waste-as-fuel activities only if they market directly to the burner.

5. Underground Injection Control: If you generate and/or treat, store, or dispose of hazardous waste, place an "X" in the box if an injection well is located at your installation. "Underground Injection" means the subsurface emplacement of fluids through a bored, drilled, or driven well; or through a dug well, where the depth of the dug well is greater than the largest surface dimension.

B) Used Oil Fuel Activities

Mark an "X" in the appropriate box(es) to indicate which used oil fuel activities are taking place at this installation.

1. Off-Specification Used Oil Fuel: If you market off-specification used oil, place an "X" in the appropriate box(es). If you burn used oil fuel place an "X" in the boxes below to indicate type(s) of combustion devices in which off-specification used oil fuel is burned. (Refer to definition section for complete description of each device.)

Note: Used oil generators are required to notify only if marketing directly to the burner.

"Other Marketer" is defined as any person, other than a generator marketing his or her used oil, who markets used oil fuel.

2. Specification Used Oil Fuel: If you are the first to claim that the used oil meets the specification established in 40 CFR § 266.40(e) and is exempt from further regulation, you must mark an "X" in this box.

Item IX -- Description of Regulated Wastes

(Only persons involved in hazardous waste activity [Item VIII.A.] need to complete this item. Transporters requesting a U.S. EPA ID Number do not need to complete this item, but must sign the certification in Item X.)

You will need to refer to 40 CFR Part 261 (enclosed as Section VII) in order to complete this section. Part 261 identifies those wastes that EPA defines as hazardous. If you need help completing this sections, please contact the appropriate addressee for your state as listed in Section III.C.

Characteristics of Nonlisted Hazardous Wastes: If you handle hazardous wastes which are not listed in 40 CFR Part 261, Subpart D, but do exhibit a characteristic of hazardous waste as defined in 40 CR Part 261, Subpart C, you should describe these wastes by the EPA hazardous waste number for the characteristic. Place an "X" in the box next to the characteristic of the wastes that you handle. If you mark "4. Toxicity Characteristic," please list the specific EPA hazardous waste number for the specific contaminants(s) in the box(es) provided. Refer to Section VIII to determine the appropriate hazardous waste number.

B) Listed Hazardous Wastes: If you handle hazardous wastes that are listed in 40 CFR Part 261, Subpart D, enter the appropriate 4-digit numbers in the boxes provided.

Note: If you handle more than 12 listed hazardous wastes, please continue listing the waste codes on an extra sheet. If an extra sheet is used, attach the additional page to the rest of the form before mailing it to the appropriate EPA Regional or State Office.

C) Other Wastes: If you handle other wastes or state regulated wastes that have a waste code, enter the appropriate code number in the boxes provided.

Item X -- Certification

This certification must be signed by the owner, operator, or an "authorized representative" of your installation. An "authorized representative" is a person responsible for the overall operation of the facility (i.e., a plant manager or superintendent, or a person of equal responsibility). *All notifications must include this certification to be complete.*

Item XI -- Comments

Use this space for any additional comments.

DEFINITIONS

The following definitions are included to help you to understand and complete the Notification Form:

ACT or RCRA means the Solid Waste Disposal Act as amended by the Resource Conservation and Recovery Act of 1976, as amended by the Hazardous and Solid Waste Amendments of 1984, 42 U.S.C. § 6901 *et seq.*

Authorized Representative means the person responsible for the overall operation of the facility or an operational unit (i.e., part of a facility), e.g., superintendent or plant manager, or person of equivalent responsibility.

Boiler means an enclosed device using controlled flame combustion and having the following characteristics:

1. the unit has physical provisions for recovering and exporting energy in the form of steam, heated fluids, or heated gases;
2. the unit's combustion chamber and primary energy recovery section(s) are of integral design (i.e., they are physically formed into one manufactured or assembled unit);

3. the unit continuously maintains an energy recovery efficiency of at least 60 percent, calculated in terms of the recovered energy compared with the thermal value of the fuel;

4. the unit exports and utilizes at least 75 percent of the recovered energy, calculated on an annual basis (excluding recovered heat used internally in the same unit, for example, to preheat fuel or combustion air or drive fans or feedwater pumps); and

5. the unit is one which the regional Administrator has determined on a case-by-case basis to be a boiler after considering the standards in 40 CFR 260.32.

Burner means the owner or operator of any boiler or industrial furnace that burns hazardous waste fuel for energy recovery and that is not regulated as a RCRA hazardous waste incinerator.

Disposal means the discharge, deposit, injection, dumping, spilling, leaking, or placing of any solid waste or hazardous waste into or on any land or water so that such solid waste or hazardous waste or any constituent thereof may enter the environment or be emitted into the air or discharged into any waters, including ground waters.

Disposal Facility means a facility or part of a facility at which hazardous waste is intentionally placed into or on any land or water, and at which waste will remain after closure.

EPA Identification (I.D.) Number means the number assigned by EPA to each generator, transporter, and treatment, storage, or disposal facility.

Facility means all contiguous land, structures, other appurtenances, and improvements on the land, used for treating, storing or disposing of hazardous waste. A facility may consist of several treatment, storage, or disposal operational units (e.g., one or more landfills, surface impoundments, or combinations of them).

Generator means any person, by site, whose act or process produces hazardous waste identified by characteristic or listed in 40 CFR Part 261.

Hazardous Waste means a hazardous waste as defined in 40 CFR § 261.3.

Hazardous Waste Fuel means a hazardous waste and any fuel that contains hazardous waste that is burned for energy recovery in a boiler or industrial furnace that is not subject to regulation as a RCRA hazardous waste incinerator. However, the following hazardous waste fuels are subject to regulation as used oil fuels:

1. Used oil fuel burned for energy recovery that is also a hazardous waste solely because it exhibits a characteristic of hazardous waste identified in Subpart C of 40 CFR Part 261; and

2. Used oil fuel mixed with hazardous wastes generated by a small quantity generator subject to 40 CFR § 261.5.

Industrial Boiler means a boiler located on the site of a facility engaged in a manufacturing process where substances are transformed into new products, including the component parts of products, by mechanical or chemical processes.

Industrial Furnace means any of the following enclosed devices that are integral components of manufacturing processes and that use controlled flame combustion to accomplish recovery of materials or energy: cement kilns, lime kilns, aggregate kilns (including asphalt kilns), phosphate kilns, coke ovens, blast furnaces, smelting furnaces, refining furnaces, titanium dioxide chloride process oxidation reactors, methane reforming furnaces, pulping liquor recovery furnaces, combustion devices used in the recovery of sulfur values from spent sulfuric acid, and other devices the Administrator may add to this list.

Marketer means a person who markets hazardous waste fuel or used oil fuel. However, the following marketers are not subject to waste-as-fuel requirements (including notification) under Subparts D and E of 40 CFR Part 266:

1. Generators and initial transporters (i.e., transporters who receive hazardous waste or used oil directly from generators including initial transporters who operate transfer stations) who do not market directly to persons who burn the fuels; and

2. Persons who market used oil fuel that meets the specification provided under 40 CFR § 266.40(e) and who are not the first to claim the oil meets the specification.

Municipality means a city, village, town, borough, county, parish, district, association, Indian tribe or authorized Indian tribal organization, designated and approved management agency under Section 208 of the Clean Water Act, or any other public body created by or under state law and having jurisdiction over disposal of sewage, industrial wastes, or other wastes.

Off-Specification Used Oil Fuel means used oil fuel that does not meet the specification provided under 40 CFR § 266.40(e).

Operator means the person responsible for the overall operation of a facility.

Owner means a person who owns a facility or part of a facility, including landowner.

Specification Used Oil Fuel means used oil fuel that meets the specification provided under 40 CFR § 266.40(e).

Storage means the holding of hazardous waste for a temporary period, at the end of which the hazardous waste is treated, disposed of , or stored elsewhere.

Transportation means the movement of hazardous waste by air, rail, highway, or water.

Transporter means a person engaged in the off-site transportation of hazardous waste by air, rail, highway, or water.

Treatment means any method, technique, or process, including neutralization, designed to change the physical, chemical, or biological character or composition of any hazardous waste so as to neutralize such waste, or so as to recover energy or material resources from the waste, or so as to render such waste non-hazardous, or less hazardous, safer to transport, store or dispose of, or amenable for recovery, amenable for storage, or reduced in volume. Such term includes any activity or processing designed to change the physical form or composition of hazardous waste so as to render it non-hazardous.

Used Oil means any oil that has been refined from crude oil, used, and as a result of such use, contaminated by physical or chemical impurities. Wastes that contain oils that have not been used (e.g., fuel oil storage tank bottom clean-out wastes) are not used oil unless they are mixed with used oil.

Used Oil Fuel means any used oil burned (or destined to be burned) for energy recovery including any fuel produced from used oil by processing, blending or other treatment, and that does not contain hazardous waste (other than that generated by a small quantity generator and exempt from regulation as hazardous waste under provisions of 40 CFR § 261.5). Used oil fuel may itself exhibit a characteristic of hazardous waste and remain subject to regulation as used oil fuel provided it is not mixed with hazardous waste.

Utility Boiler means a boiler that is used to produce electricity, steam, or heated or cooled air, or other gases or fluids for sale.

Waste Fuel means hazardous waste fuel or off-specification used oil fuel.

ALPHABETIZED STATE LISTING OF HAZARDOUS WASTE CONTACTS

Alabama

Land Division
Alabama Department of Environmental Management
1751 Federal Drive
Montgomery, AL 36130
(205) 271-7730

Alaska

EPA Region X
Waste Management Branch
MS HW-112
1200 Sixth Avenue
Seattle, WA 98101
(206) 442-0151

American Samoa

Environmental Quality Commission
Government American Samoa
Pago Pago, American Samoa 96799
(Commercial Call (684) 633-4116)
Country Code 663-2304

Arizona

Office of Waste and Water Quality Management
Arizona Department of Environmental Quality
2005 N. Central Ave., Room 304
Phoenix, AZ 85004
(602) 257-2305

Arkansas

Arkansas Department of Pollution Control and Ecology
P.O. Box 9583
Little Rock, AR 72219
(501) 562-7444

California

California Department of Health Services
Toxic Substances Control Division
Department of Health Services
P.O. Box 942732, 400 P. Street
Sacramento, CA 95814
(916) 323-2913

Colorado

Colorado Department of Health
Waste Management Division
4300 Cherry Creek Drive So.
Denver, CO 00222-1530
(303) 692-2000

Connecticut

Waste Management Bureau
Department of Environmental Protection
State Office Building
Hartford, CT 06016
(203) 566-8844

Delaware

Delaware Department of Natural Resources and Environmental Control
Division of Air and Waste Management
Hazardous Waste Management Branch
P.O. Box 1401, 89 Kings Highway
Dover, DE 19903
(302) 736-3689

District of Columbia

Department of Consumer and Regulatory Affairs
Environmental Control Division
Pesticides and Hazardous Waste Branch
2100 Martin Luther King, Jr. Ave, S.E.
Room 204
Washington, DC 20020
(202) 727-7000

Florida

Hazardous Waste Section
Department of Environmental Regulations
Twin Towers Office Bldg.
2600 Blair Stone Road
Tallahassee, FL 32399-2400
(904) 488-0300

Georgia
Land Protection Branch
Industrial and Hazardous Waste Management Program
Floyd Towers East 205 Butler Street, S.E.
Atlanta, GA 30334
(404) 656-2833

Guam

Guam EPA
IT&E
Harmon Plaza Complex, Unit D-107
130 Rojas St.
Harmon, Guam 96911
(Overseas Operator) (671) 646-7579

Hawaii

To obtain information or forms contact:

Hawaii Department of Health
Hazardous Waste Program
P.O. Box 3378
Honolulu, HI 96801
(808) 548-2270

Mail completed forms to:

U.S. EPA Region IX
RCRA Programs Section
Hazardous Waste Management Division
1235 Mission Street
San Francisco, CA 94103

Idaho

Idaho Department of Health & Welfare
Tower Building, Third Floor
450 West State Street
Boise, ID 83720
(208) 334-5879

Illinois

To obtain information or forms contact:

U.S. EPA Region V
RCRA Activities
Waste Management Division
77 Jackson Street
Chicago, IL 60604
(312) 886-4001

Mail your completed forms to:

Illinois Environmental Protection Agency
Division of Land Pollution Control
2200 Churchill Road
Springfield, IL 62706
(217) 782-6760

Indiana

To obtain information or forms contact:

Indiana Dept. of Environmental Management
105 S. Meridian Street
P.O. Box 6015
Indianapolis, IN 46225
(317) 323-3210

Mail your completed forms to:

U.S. EPA Region V
RCRA Activities
Waste Management Division
77 Jackson Street
Chicago, IL 60604

Iowa

U.S. EPA Region VII
RCRA Branch
726 Minnesota Avenue
Kansas City, KS 66101
1-800-223-0425 or
(913) 236-2852

Kansas

Bureau of Air and Waste Management
Department of Health and Environment
Forbes Field, Bldg. 740
Topeka, KS 66620
(913) 296-1600

Kentucky

Division of Waste Management
Department for Environmental Protection
Cabinet for Natural Resources
& Environmental Protection
Fort Boone Plaza, Building No. 2
Frankfort, KY 40601
(502) 564-6716

Louisiana*

Louisiana Department of Environmental Quality
Department of Solid and Hazardous Waste
P.O. Box 44307
Baton Rouge, LA 70804
(504) 342 -1354

**In Louisiana you must have an EPA ID Number*

Maine

Bureau of Oil and Hazardous Materials Control
Department of Environmental Protection
Ray Bldg. Station #17
Augusta, ME 04333
(207) 289-2651

Maryland

Maryland Department of the Environment
Waste Management Administration
2500 Broening Highway
Baltimore, MD 21224
(301) 631-3304

Massachusetts

Division of Hazardous Waste
Department of Environmental Protection
One Winter Street, 5th Floor
Boston, MA 02108
(617) 292-5851

Michigan

To obtain information or forms contact:

Waste Management Division
Environmental Protection Bureau
Department of Natural Resources
Box 30038
Lansing, MI 48909
(517) 373-2730

Mail your completed forms to:

U.S. EPA Region V
RCRA Activities
Waste Management Division
P.O. Box A3587
Chicago, IL 60690

Minnesota

To obtain information or forms contact:

Solid and Hazardous Waste Division
Minnesota Pollution Control Agency
520 Lafayette Road, North
St. Paul, MN 55155
(612) 296-7282

Mail your completed forms to:

U.S. EPA Region V
RCRA Activities
Waste Management Division
P.O. Box A3587
Chicago, IL 60690

Mississippi

Hazardous Waste Division
Bureau of Pollution Control
P.O. Box 10385
Jackson, MS 39289-0385
(601) 961-5062

Missouri

Waste Management Program
Department of Natural Resources
Jefferson Building
205 Jefferson Street, (13/14 floor)
P.O. Box 176
Jefferson City, MO 65102
(314) 751-3176

Montana

Solid and Hazardous Waste Bureau
Department of Health and Environmental Sciences
Cogswell Building
Helena, MT 59620
(406) 444-1430

Nebraska

Hazardous Waste Management Section
Department of Environmental Control
P.O. Box 98922
Lincoln, NE 68509-8922
(402) 471-4217

Nevada

Waste Management Bureau
Division of Environmental Protection
Department of Conservation & Natural Resources
Capitol Complex
123 West Nye Lane
Carson City, NV 89710
(702) 687-5872

New Hampshire

Department of Environmental Services
Waste Management Division
6 Hazen Drive
Concord, NH 03301
(603) 271-2900

New Jersey

To obtain information or forms contact:

New Jersey Department of Environmental Protection
Bureau of Hazardous Waste Classification and Manifests
401 East State Street, CN-028
Trenton, NJ 08625
(609) 292-8341

Mail your completed forms to:

U.S. EPA Region II
Permits Administration Branch
26 Federal Plaza, Room 505
New York, NY 10278

New Mexico

New Mexico Health & Environmental Dept.
Hazardous Waste Bureau
1190 St. Francis Drive
Santa Fe, NM 87503
(505) 827-2929

New York

To obtain information contact:

New York Department of Environmental Conservation
Division of Hazardous Waste Substances Regulation
P.O. Box 12820
Albany, NY 12212
(518) 457-0530

Mail your completed forms to and obtain forms from

U.S. EPA Region II
Permits Administration Branch
26 Federal Plaza
New York, NY 10278

North Carolina

Hazardous Waste Section
Division of Soid Waste Management
Department of Environment, Health and Natural Resources
P.O. Box 27687
Raleigh, NC 27611-7687
(919) 733-2178

North Dakota

Division of Waste Management
Department of Health and Consolidated Labs.
1200 Missouri Ave.
P.O. Box 5520
Bismarck, ND 58502-5520
(701) 224-2366

North Mariana Islands, Commonwealth of

To obtain information or forms contact:

Department of Public Health and Environmental Services
Division of Environmental Quality
Dr. Torres Hospital
P.O. Box 1304
Saipan, Mariana Islands 96950
Overseas Operator: (676) 234-6984
Cable address: GOV. NMI Saipan

Mail completed forms to:

U.S. EPA Region IX
RCRA Programs Section (H-2-3)
Hazardous Waste Management Division
1235 Mission Street
San Francisco, CA 94103

Ohio

U.S. EPA Region V
RCRA Activities
Waste Management Division
77 Jackson Street
Chicago, IL 60604
(312) 886-4001

Oklahoma

Oklahoma State Department of Health
Industrial Waste Division
1000 Northeast 10th Street
Oklahoma City, OK 73152
(405) 271-5338

Oregon

Oregon Department of Environmental Quality
Hazardous Waste Operations
811 Southwest 6th Avenue
Portland, OR 97204
(503) 229-5913

Pennsylvania
To obtain information or forms contact:

Pennsylvania Department of Environmental Resources
Bureau of Waste Management
P.O. Box 2063
Harrisburg, PA 17120
(717) 787-9870

Mail your completed forms to:

U.S. EPA Region III
RCRA Programs Branch
Pennsylvania Section (3 HW51)
841 Chesnut Building
Philadelphia, PA 19107

Puerto Rico

To obtain information or forms contact:

Environmental Quality Board
Land Pollution Control Area
P.O. Box 11488
Santurce, PR 00910-1488
(809) 722-0439

Mail your completed forms to:

U.S. EPA Region II
Permits Administration Branch
26 Federal Plaza, Room 505
New York, NY 10278

Rhode Island

Division of Air and Hazardous Materials
Department of Environmental Management
291 Promenade Street
Providence, RI 02908-5767
(401) 277-2808

South Carolina

Bureau of Solid Waste Management
Hazardous Waste Management
Department of Health and Environmental Control
2600 Bull Street
Columbia, SC 29201
(803) 734-2500

South Dakota

Office of Waste Management
South Dakota Department of Water and Natural Resources
Joe Foss Building, 523 East Capitol Street
Pierre, SD 57501-3181
(605) 773-3153

Tennessee

Division of Solid Waste Management
Tennessee Department of Health and Environment
Customs House, 4th Floor
701 Broadway
Nashville, TN 37247-3530
(615) 741-3424

Texas

Texas Water Commission
Compliance Assistance Unit
Hazardous and Solid Waste Division
P.O. Box 13087, Capitol Station
Austin, TX 78711-3087
(512) 463-8175

Utah

Bureau of Solid and Hazardous Waste
Department of Health
P.O. Box 16690
288 North 1460 West
Salt Lake City, UT 84116-0690
(801) 538-6170

Vermont

Hazardous Materials Management Division
Department of Environmental Conservation
103 South Main Street
Waterbury, VT 05676
(802) 244-8702

Virgin Islands

To obtain information or forms contact:

Department of Planning & Natural Resources
Division of Environmental Protection
179 Altona and Welgunst
St. Thomas, VI 00801
(809) 774-3320

Mail your completed forms to :

U.S. EPA Region II
Permits Administration Branch
26 Federal Plaza, Room 505
New York, NY 10278

Virginia

Virginia Department of Waste Management
Monroe Building, 11th Floor
101 North 14th Street
Richmond, VA 23219
(804) 225-2667

Washington

Solid and Hazardous Waste Management Division
Department of Ecology, Mail Stop PV-11
Olympia, WA 98504
(206) 459-6369

West Virginia

West Virginia Department of Natural Resources
Waste Management Section
1356 Hansford Street
Charleston, WV 25301
(304) 384-5393

Wisconsin

To obtain information or forms contact:

Bureau of Solid Waste
Department of Natural Resources
P.O. Box 7921
Madison, WI 53707
(608) 266-1327

Mail your completed forms to:

U.S. EPA Region V
RCRA Activities
Waste Management Division
77 Jackson Street
Chicago, IL 60604

Wyoming

EPA Region VIII
Waste Management Division (8HWM-0N)
999 18th Street, Suite 500
Denver, CO 80202-2405
(303) 293-1795

US EPA Regional Offices

Region	Geographical Area Covered	EPA Regional Office
I	Connecticut, Maine, Massachusetts, New Hampshire, Rhode Island, Vermont	US EPA Region I State Waste Programs Branch JFK Federal Building Boston, MA 02203-2211
II	New Jersey, New York, Puerto Rico, Virgin Islands	US EPA Region II Air and Waste Management Division 26 Federal Plaza New York, NY 10278
III	Delaware, District of Columbia, Maryland, Pennsylvania, Virginia, West Virginia	US EPA Region III Waste Management Branch MS 3 HW 34 841 Chestnut Street Philadelphia, PA 19107
IV	Alabama, Florida, Georgia, Kentucky, Mississippi, North Carolina, South Carolina, Tennessee	US EPA Region IV Hazardous Waste Management Division 345 Courtland Street, NE Atlanta, GA 30365
V	Illinois, Indiana, Michigan, Minnesota, Ohio, Wisconsin	US EPA Region V RCRA Activities Waste Management Division 77 Jackson Street Chicago, IL 60690
VI	Arkansas, Louisiana, New Mexico, Oklahoma, Texas	US EPA Region VI Hazardous Waste Materials Division First Interstate Bank Tower 1445 Ross Avenue, Suite #1200 Dallas, TX 75202-2733
VII	Iowa, Kansas, Nebraska, Missouri	US EPA Region VII RCRA Branch 726 Minnesota Avenue Kansas City, KS 66620
VIII	Colorado, Montana, North Dakota, South Dakota, Utah, Wyoming	US EPA Region VIII Waste Management Division (8HWM-ON) 999 18th Street, Suite 500 Denver, CO 80202-2405
IX	Arizona, California, Hawaii, Nevada, American Samoa, Guam	US EPA Region IX Hazardous Waste Management Division 1235 Mission Street San Francisco, CA 94103
X	Alaska, Idaho, Oregon, Washington	US EPA Region X Waste Management Branch (MW-112) 1200 Sixth Avenue Seattle, WA 98101

CHAPTER 5

IDENTIFYING HAZARDOUS WASTE, OR MAKING THE HAZARDOUS WASTE DECISION

It is extremely important that the RCRA administrator identify hazardous waste generating operations. It is, however, unlikely that the facility generators will have the expertise to decide whether their waste, in each and every case, is or is not regulated. The instructions provided in this chapter are basic and will allow the generator to identify ninety-plus percent of his waste as either regulated or nonregulated. It may be necessary for the RCRA administrator to make the decision on the remaining ten percent, or employ additional internal staff or retain the services of an attorney or consultant. The following chapter will give instructions on completing a hazardous waste profile for each waste stream the facility generators produce, which will provide the necessary information for making a definitive decision as to the regulatory status of the waste stream.

IDENTIFYING A HAZARDOUS WASTE

The generally held concept --that a "hazardous waste" is an acutely dangerous substance that is also a waste-- is much narrower than the definition promulgated by the Environmental Protection Agency (EPA). It will be necessary to educate your generators on what is considered a RCRA hazardous waste. This can be accomplished by reviewing the spirit of the statutory definition during Waste Generator Training. It is important that generators complete a hazardous waste characterization and profiling form for each waste stream generated. Usually, the company will employ a consultant or train a small internal staff to accomplish the preliminary profiling.

DEFINITION OF A HAZARDOUS WASTE

The following is a preliminary training for management and staff for use in determining what is a hazardous waste:

To be a hazardous waste, the waste must first be a solid waste. The definition of a solid waste (40 CFR § 261.2) includes any solid, semi-solid, or contained gas which is either discarded, has served its intended purpose, or is a manufacturing or mining by-product. The definition is very inclusive. The EPA specifically excludes certain wastes from regulations under RCRA.

The following are excluded: domestic sewage, Clean Water Act point source discharges (i.e., facilities that have National Pollution Discharge Elimination Permits), irrigation return flows, *in situ* mining waste or Atomic Energy Act source, special nuclear or by-product material and other specific identified wastes.

Remember if a material is a waste but is listed in one of the RCRA exclusions, it is not a RCRA solid waste. For a more in-depth look at exclusion, review 40 CFR § 261.4.

Next, one must determine if the solid waste meets the EPA definition of a hazardous waste. This is determined by "listing" or "characteristic." If a waste is listed, it is automatically considered hazardous, and is specifically described and given a waste code. Listed wastes make up the "F", "K", "P" and "U" tables.

The "F" wastes are from non-specific sources (40 CFR § 261.31). The "K" wastes are from specific sources or processes (40 CFR § 261.32). The "P" wastes are all-specific acutely hazardous wastes, which are commercially pure grades or technical grades of chemicals which constitute the sole active ingredient in a waste (40 CFR § 261.33). The "U" wastes are analogous to the "P" waste, except that they are not acutely hazardous wastes.

Some examples of each category are listed below:

1. F001 - The following spent halogenated solvents used in degreasing: tetrachloroethylene, trichloroethylene, methylene chloride, 1,1,1-trichloroethane, carbon tetrachloride, and chlorinated fluorocarbons; all spent solvent mixtures/blends used in degreasing containing, before use, a total of ten percent or more (by volume) of one or more of the above halogenated solvents or those solvents listed in F002, F004, and F005; and still bottoms from the recovery of these spent solvents and spent solvent mixtures.

2. F007 - spent cyanide plating solution from electroplating operations.
3. K001 - Bottom sediment sludge from the treatment of wastewater from wood preserving processes that use creosote and/or pentachlorophenol.
4. P005 - Allyl alcohol.
5. U001 - Acetaldehyde.

In summary, if a solid waste is not excluded and is listed, then it is a RCRA Hazardous Waste.

The second way a solid waste may be defined as a RCRA Hazardous Waste is if it exhibits any one or more of the following four characteristics; ignitability (40 CFR § 261.21), corrosivity (40 CFR § 261.22), reactivity (40 CFR § 261.23) and toxicity (40 CFR § 261.24).

A waste exhibits the characteristic of ignitability if a representative sample has any of the following properties: (1) it is a liquid and has a flash point (closed cup) of less than 140 ^{0}F (60^{0}C); (2) it is a non-liquid and may spontaneously ignite and burn vigorously and persistently; (3) it is a DOT defined ignitable compressed gas; or (4) it is a DOT defined oxidizer. A solid waste that exhibits the characteristics of ignitability and is not listed has the Hazardous Waste Number of D001.

A waste exhibits the characteristic of corrosivity if a representative sample has any of the following properties: (1) it has a pH less than or equal to 2, or greater than or equal to 12.5 or (2) it is a liquid and corrodes mild steel at rates greater than 0.25 inches (6.35 mm) per year. A solid waste that exhibits the characteristics of corrosivity and is not listed has the Hazardous Waste Number of D002.

A waste exhibits the characteristic of reactivity if a representative sample has any of the following properties: (1) is normally unstable; (2) reacts visibly or explosively with water; (3) forms potentially explosive mixture with water; (4) generates toxic gas or vapor when mixed with water; (5) is a cyanide or sulfide bearing waste that can generate toxic gas when exposed to pH conditions between 2 and 12.5; (6) is explosive; or (7) is a DOT class A or B explosive. A solid waste that exhibits the characteristics of reactivity and is not listed has the Hazardous Waste Number of D003.

A waste exhibits the characteristic of toxicity if using the TCLP test (Toxicity Characteristic Leaching Procedure), a representative sample extract exceeds the maximum concentration limit for the contaminant.

TCLP test: An acetic acid solution is added to the solids from a sample, and the resultant mixture is tumbled for 18 hours at 22 +/- 3 ^{0}C in a zero headspace extractor which prevents volatilization of certain constituents. The liquids are expressed out of the container and collected in an airtight bag. An aliquot of the liquid is introduced into a purge and trap device where helium is bubbled through the liquid to remove the volatile fractions which are concentrated on a Tenax Trap. The trap is then heated, and the volatile organics are swept or flushed into a gas chromatograph/mass spectrometer (GC/MS). The semi-volatiles in the liquid go through a sample extraction/preparation stage and then are directly injected into the GC/MS. The final values are then compared to the regulatory thresholds to determine whether the waste should be classified as hazardous. (For the 40 chemicals TCLP has added to the list of RCRA-regulated chemicals, see table on page 50, Toxicity Characteristic Constituents and Regulatory Levels.)

Toxicity Characteristic Constituents and Regulatory Levels

EPA #(1)	Constituent	Reg. Level (non-wastewater) (mg/L)
D004	Arsenic	5.00
D005	Barium	100.00
D018	Benzene	.50
D006	Cadmium	1.00
D019	Carbon Tetrachloride	.50
D020	Chlordane	.03
D021	Chlorobenzene	100.00
D022	Chloroform	6.00
D007	Chromium	5.00
D023(3)	o-Cresol	200.00
D024(3)	m-Cresol	200.00
D025(3)	p-Cresol	200.00
D026(3)	Cresol	200.00
D016	2,4-D	10.00
D027	1,4-Dichlorobenzene	7.50
D028	1,2-Dichloroethane	.50
D029	1,1-Dichloroethylene	.70
D030(2)	2,4-Dinitrotoluene	.13
D012	Endrin	.02
D031	Heptachlor (and its hydroxide)	.008
D032(2)	Hexachlorobenzene	.13
D033	Hexachloro-1,3-butadiene	.50
D034	Hexachloroethane	3.00
D008	Lead	5.00
D013	Lindane	.40
D009	Mercury	.20
D014	Methoxychlor	10.00
D035	Methyl Ethyl Ketone	200.00
D036	Nitrobenzene	2.00
D037	Pentachlorophenol	100.00
D038(2)	Pyridine	5.00
D010	Selenium	1.00
D011	Silver	5.00
D039	Tetrachloroethylene	.70
D015	Toxaphene	.50
D040	Trichloroethylene	.50
D041	2,4,5-Trichlorophenol	400.00
D042	2,4,6-Trichlorophenol	2.00
D017	2,4,5-Trichlorophenoxy-propionic Acid (Silvex)	1.00
D043	Vinyl Chloride	.20

1. Hazardous Waste Number
2. Quantification limit is greater than the calculated regulatory level. The quantification limit therefore becomes the regulatory level.
3. If o-, m- and p-cresol concentrations cannot be differentiated, the total cresol (D026) concentration is used. The regulatory level for total cresol is 200 mg/L.

Please note that the EPA has recently required the use of data correction to account for the recoveries of sample spikes. This change in data calculation may cause more waste to become subject to the hazardous waste regulations.

To recap, first check the tables titled "40 CFR Subpart D - List of Hazardous Waste" to see if the solid waste is listed. If it is listed, then use the assigned hazardous waste number. If the solid waste is not listed but exhibits one of the four characteristics, then use the appropriate hazardous waste number.

The above sounds simple, but in reality this determination is much more complex. The regulation allows the generator to use process knowledge to determine if a solid waste is a hazardous waste. Process knowledge is much less expensive than doing analytical tests. But if process knowledge is lacking or is incomplete and no listed waste code number is appropriate, tests for characteristics will have to be conducted.

As we will see later, it is important to have the correct waste code number(s) to determine the disposal method.

INFORMATION NECESSARY FOR CHARACTERIZING WASTE

Suggested questions and information needed for characterizing solid waste and determining if the waste is hazardous.

- Describe the process that generates the waste.
- What is the physical form of the waste. (Solid, liquid, gas, sludge, powder; is liquid single or multi-phase?)
- Do you have MSDSs for constituents or process chemicals incorporated into the waste?
- If yes to the above question, list those chemicals.
- Are any other hazardous substances or materials incorporated into the waste?
- From the above information can you determine if it is listed waste? If yes, list the appropriate waste number.
- If the waste is not listed continue checking for hazardous characteristics.
- If the waste is an organic liquid, what is the flash point? If yes, it is less than 140 °F, the waste is an ignitable (D001). Does it contain halogenated organics? If the answer is yes, it is a possible F001-3.

- Does it react with air or water? If so, it may be a reactive waste (D003).
- If the waste is an aqueous liquid, check the pH. If it is less than or equal to 2, or greater than or equal to 12.5, then it is a corrosive with waste code D002.
- If it is a liquid or solid suspected of containing heavy metals, have the waste analysis for 8 RCRA metals. If you use a total analysis, use the 20 x rule for assumed TCLP levels (for solids, divide total metal concentration by 20 and use resulting number for assumed TCLP lower limit). If this waste is close to (80% or greater) the regulatory limit, analyze the sample using the TCLP protocol.
- If it is a liquid or solid suspected of containing regulated organics, conduct an organic TCLP analysis.

THE NEED FOR WASTE PROFILING

This chapter has shown that determining whether your waste is hazardous, and determining its characteristics, can be an involved process. It is important that each waste stream to be profiled use a standardized system in determining whether your waste is hazardous and how it should be regulated and disposed. The next chapter provides such a standardized system for profiling waste streams.

CHAPTER 6

CONDUCTING HAZARDOUS WASTE PROFILING

It is extremely important that the RCRA administrator have his generators complete a hazardous waste profile for each waste stream they produce. The profile will be the backbone of a system for tracking, storage and disposal of hazardous waste.

It is unlikely that the facility generators will have the expertise to complete the profile. For a smaller operation, it may be incumbent on the RCRA administrator to complete the profile. Other options include training internal staff or retaining the services of a consultant.

STANDARD PROCEDURES FOR CONDUCTING WASTE PROFILING

Arrange an exploratory (an initial scoping) meeting with the generator to review the profiling process. Explain the waste profiling process to the generator. Request and record the generator's group charge number (if used) that will be used to track profile costs.

The following preliminary data should be recorded either on a one- or two-page summary report, or a waste profiling form:

- Date of meeting
- Technical contact's name
- Technical contact's phone number
- Generator group's name
- Generator group's physical office location (Not P.O. Box)

The available analytical information should be reviewed to determine if it is sufficient for characterizing the waste stream. A decision will be made by the profiling personnel whether to conduct further analytical analysis or to go to the next step. If any chemical analyses are needed to characterize the waste, the profiling personnel should specify the type of analysis to be performed and assist the generator in contacting appropriate facilities. The profiling personnel should provide the generator with the needed analysis request forms (pages 70-73) and complete the forms with the generator.

After the initial meeting a more in-depth meeting should be conducted. This meeting may immediately follow the scoping meeting as appropriate. This meeting should include personnel who have in-depth knowledge of the processes that generate the waste stream(s). They should conduct a discussion of the process that generates the waste with appropriate questions directed toward determining if the waste fits the definition of a listed waste. An appropriate waste profile number should be assigned to each identified waste stream.

The following information should be recorded / filed:

- The agreed-upon name of each waste stream.
- A synopsis statement that describes the process that generates the waste.
- Copies of available material safety data sheets (MSDS) for the chemical constituent of the waste stream.
- A copy of any analytical analysis conducted on the waste(s).
- The waste generation rate.

Other known physical and chemical characteristics of the waste should be recorded. Examples are:

- Waste form

- pH
- Flashpoint
- Percent water
- Trace contaminants

The profiling personnel should have obtained all necessary information from the generator so that he can complete the profile form. At least two copies of the waste profile packet should be prepared, one for the hazardous waste management facilities' use and one for the generator.

A completed waste profile packet should contain:

- Profile explanation cover sheet.
- Completed waste profile form.
- Land Disposal Restriction form.
- California list form if appropriate.
- Example bill of lading or hazardous waste manifest as appropriate.
- Example hazardous or non-hazardous labels and marking as appropriate.
- A copy of appropriate DOT guide page.
- Profile description information form.
- One copy of HWMF acceptance and/or rejection form.

A meeting should be held with the manager of the permitted storage facility to review profiles for mistakes, omissions, and to obtain his approval.

The manager should:

- Accept or not accept the profile, or request changes or additions.
- If the profile is approved, the current date should be recorded on all copies of the form.
- Determine whether the waste stream can be accepted for storage or treatment at the facility.
- Select the appropriate generator notice letter to send to the generator and complete the necessary sections on all copies of the letter.
- File one copy of the profile at the facility after the generator has signed the profile.

A meeting should be held with the generator to review the profile. The generator should sign and date all copies of the profiles that are complete and accurate.

The RCRA administrator's computerized profiling systems should be updated with information from the new profile. The master profile list should be updated and an archive copy of the file made. Periodic master profile lists should be supplied to the facility manager.

SAMPLE HAZARDOUS WASTE MANAGEMENT FACILITY UNIFORM WASTE PROFILE FORM

PAGE 1
HWMF PROFILE NO.:____________
PROFILE APPROVAL DATE: ___/___/______.
ACCEPTABLE FOR STORAGE AT HWMF (Y,N): ____ACTIVE(Y,N):_______
LETTER SENT (Y,N):_______
PROFILE RENEWAL DATE NOTIFICATION (APPROVAL DATE + 1YR - 60 DAYS):___/___/_____.
PROFILE ANNUAL RECERTIFICATION DATE:___/___/_____.
OFFSITE TSD:____________OFFSITE TSD PROFILE DATE:___/___/_____.
DISPOSAL PROFILE NUMBER:______________________________

A. GENERATOR INFORMATION

1. Generator Facility: ______________________________
2. Technical Contact: ______________________________
3. Phone No.: (   )______________________________
4. Emergency Contact, Agency: ______________________________
5. Emergency Contact, Person: ______________________________
6. Phone No.: (   ) ______________________________
7. Generator US EPA ID: ______________________________
8. Generator ID Number: ______________________________
9. Generator Cost Code: ______________________________

B. PICKUP LOCATION

1. Company Name: ______________________________
2. Group or Division: ______________________________
3. Address: ______________________________
4. City: ______________________________ 5. State: ______ 6. Zip:________
7. Pickup Contact:______________________________ 8. Phone No.: (   )__________

C. GENERAL WASTE DESCRIPTION

1. Name of Waste:______________________________
2. Process Generating Waste: ______________________________
3. Quantity Generated:____________________ drums:_______pounds:_______gallons:_______
4. Period: ______________________________

PAGE 2 HWMF PROFILE NO.:

D. WASTE COMPOSITION (Must add up to a minimum of 100%)

Chemical Name & Constituent Print Chemical Root (Comma) Prefix :	: Concentration. Min %	Max.%	mg/L:	: CASRN(1)	: RQ(2) : lbs
1)					
2)					
3)					
4)					
5)					
6)					
7)					
8)					
9)					
10)					
11)					
12)					
13)					
14)					
15)					
16)					
17)					
18)					
19)					
20)					

(1) Chemical Abstract Registry Number
(2) Reportable Quantity

PAGE 3 HWMF PROFILE NO.:

E. PROPERTIES OF WASTE (N/A for those which do not apply)

1. Color:______________________ 2. Odor:______________________

3. Ignitability (Flash Point):__________ °F (For Liquids)

4. pH ________or Corrosivity ________mm/yr. (For Liquids)

5. Melting Point: ______ °F (For Solids)

6. Boiling Point: ______ °F (For Liquids)

7. Specific Gravity (Liquids):__________or Bulk (Solids):_______lb./cu. ft.

8. Viscosity__________CPS @ 25° C (77 °F) (For Liquids)

9. Vapor Pressure:______mmHg @ ______ °C (For Liquids)

10. Heat of Combustion:______________BTU/lb (For Liquids and Solids)

11. Bromine:__________% 12. Iodine:________________%

13. Fluorine:__________% 14. Chlorine:________________%

15. Sulfur:___________% 16. Flowable at 25 °C (77 F) (Y, N):_____

17. Physical state at 25 °C (77 °F) and phase by vol.
(check all that apply) (No. of phases)

Powder	________%	____________
Gas	________%	____________
Sludge/Semisolid	________%	____________
Liquid	________%	____________
Solid	________%	____________
Other	________%	____________

18. Percent Solid by Weight: ____________% (For Solids & Sludges)
Percent Solids (Sludge): ____________% (For Liquids)

19. TOC (Total Organic Carbon): ____________ mg/L (For Liquids)
Method:

20. VOC (Volatile Organic Compounds):________mg/L (For Liquids)

21. Halogenated Organic Compounds: __________% (For Solids & Liquids) or 22.

22. Halogenated Organic Compounds: __________ mg/L, mg/Kg (For Solids & Liquids)
Method:

23. Cyanides: _________mg/L, mg/Kg (For Solids and Liquids)

PAGE 4 HWMF PROFILE NO.:

F. ELEMENTAL ANALYSIS (Total Metals and TCLP metals are required for incineration waste streams. TCLP metals data are required for landfill material. For liquid wastes with less than 0.5% solids: Total metals constituents expressed in mg/kg multiplied by the density of waste (in kg/L) are equal to the TCLP results (in mg/L).

Total Metals: (Unit Type: mg/kg, mg/L, ppm)

Antimony (Sb)____________________
Arsenic (As)____________________
Barium (Ba) ____________________
Beryllium (Be) ____________________
Cadmium (Cd) ____________________
Chromium (Cr) ____________________
(CrVI)____________________
Cobalt (Co) ____________________
Lead (Pb)____________________
Manganese (Mn)____________________
Mercury (Hg) ____________________
Molybdenum (Mo)____________________
Nickel (Ni)____________________
Potassium (K) ____________________
Selenium (Se) ____________________
Silicon (Si) ____________________
Silver (Ag Blend) ____________________
Sodium (Na) ____________________
Thallium (Th) ____________________
Vanadium (V) ____________________
Zinc (Zn) ____________________

TCLP	(mg/L)	REGULATED TC LEVEL	TCLP	(mg/L)	REGULATED TC LEVEL
D004-Arsenic		5.0	D008-Lead		5.0
D005-Barium		100.0	D009-Mercury		0.2
D006-Cadmium		1.0	D010-Selenium		1.0
D007-Chromium		5.0	D011-Silver		5.0

RCRA Metals Levels Found by:
TCLP Analysis (Y, N):__
Calculation from Total Analysis (Y, N):___
Process Knowledge (Y, N):______

PAGE 5 HWMF PROFILE NO.:

G. ORGANIC TOXIC CHARACTERISTIC WASTE

:EPA CONTAMINANT	REGULATED CONC.(MG/L)	TOTAL ANALYSIS	OR TCLP ANALYSIS
:D012 ENDRIN	0.02		
:D013 LINDANE	0.4		
:D014 METHOXYCHLOR	10.0		
:D015 TOXAPHENE	1.0		
:D016 2,4-D	10.0		
:D017 2,4,5-TP (SILVEX)	1.0		
:D018 BENZENE	0.5		
:D019 CARBON TETRACHLORIDE	0.5		
:D020 CHLORDANE	0.03		
:D021 CHLOROBENZENE	100.0		
:D022 CHLOROFORM	6.0		
:D023 o-CRESOL	200.0		
:D024 m-CRESOL	200.0		
:D025 p-CRESOL	200.0		
:D026 CRESOL	200.0		
:D027 1,4-DICHLOROBENZENE	7.5		
:D028 1,2-DICHLOROETHANE	0.5		
:D029 1.1-DICHLOROETHYLENE	0.7		
:D030 2,4-DINITROTOLUENE	0.13		
:D031 HEPTACHLOR (& its hydroxides)	0.008		
:D032 HEXACHLOROBENZENE	0.13		
:D033 HEXACHLOROBUTADIENE	0.5		
:D034 HEXACHLOROETHANE	3.0		
:D035 CARBON TETRACHLORIDE	0.5		
:D036 METHYL ETHYL KETONE	200.0		
:D037 PENTACHLOROPHENOL	100.0		
:D038 PYRIDINE	5.0		
:D039 TETRACHLOROETHYLENE	0.7		
:D040 TRICHLOROETHYLENE	0.5		
:D041 2,4,5-TRICHLOROPHENOL	400.0		
:D042 2,4,6-TRICHLOROPHENAL	2.0		
:D043 VINYL CHLORIDE	0.2		

Organic TCLP Levels Found by:
TCLP Analysis (Y, N):___
Calculation from Total Analysis (Y, N):___
Process Knowledge (Y, N):______

PAGE 6 HWMF PROFILE NO.:

H. REACTIVE CHARACTERISTICS (Mark all Y, N)

1. Explosive:__________
2. Pyrophoric:_________
3. Shock Sensitive:______
4. Water Reactive:______
5. Air Reactive:________
6. Reactive Cyanide:____(mg/kg)________
7. Reactive Sulfide:_____(mg/kg)________
8. Oxidizers: (specify):____
9. Other: (specify):____

I. REGULATORY COMPLIANCE (If present enter maximum conc.)

1. OSHA Carcinogens ppm

OSHA Carcinogens	ppm
1. Acrylonitrile (Vinyl Cyanide)	____________
2. 2-Acetylaminofluorene	____________
3. Aminodiphenyl	____________
4. Asbestos	____________
5. Benzidine	____________
6. Benzene	____________
7. bis-Chloromethyl Ether	____________
8. 3,3'-Dichlorobenzidine (and salt)	____________
9. Dimethylaminoazobenzene	____________
10. 1,2-Dibromo-3-Chloropropane	____________
11. Ethylenimine	____________
12. Ethylene Oxide	____________
13. Formaldehyde	____________
14. Methyl Chloromethyl Ether	____________
15. 4,4'-Methylene bis (2-Chloroaniline)	____________
16. alpha-Napthylamine	____________
17. beta-Napthylamine	____________
18. 4-Nitro Biphenyl	____________
19. N-Nitrosodimethylamine	____________
20. beta-Propiolactone	____________
21. Vinyl Chloride	____________

2. Radioactive Material (Y,N):___

3. TSCA Regulated PCB (Y,N):________Concentration:__________ppm

4. Biomedical Waste (Y,N):_____Infectious Waste (Y,N):_________

5. FIFRA Pesticides with Specific Disposal Requirements (Y,N):____

6. Asbestos (Y,N):______

7. California List Regulated Waste (Y,N):_____

8. Solvent Regulated Waste (Y,N):______

9. Dioxin Regulated Waste (Y,N):______

PAGE 7 HWMF PROFILE NO.:

J. SHIPPING INFORMATION

1. EPA RCRA Hazardous Waste (Y, N):______________________

2. Wastewater (Y, N):______________________

3. All applicable EPA Hazardous Waste Codes and subcategories including characteristic ('D') codes:

____

4. DOT Regulated Material (Y, N):______________________

5. DOT Shipping Name: ______________________

6. Additional Information:______________________

7. DOT Hazard Class: ______________________

8. DOT (UN/NA) No.:______________ 9. Packing Group No.: ________

10. DOT Guide No.:______________ 11. RQ (lbs.):______________

12. RQ (Substance):______________________

13. EPA Compatibility Group:______________________

K. MANIFEST INFORMATION/SURPLUS DISTRIBUTION AND DELIVERY REPORTS

1. Company Name:______________________

2. Address:______________________

3. City:______________ 4. State_______ 5. Zip:______________

6. Manifest Contact:______________ 7. Phone: ( ) ______

L. METHOD OF SHIPMENT

1. Drums (Y,N):___1(a). Number of drums:______________
1(b). Drums Metal (Steel) DOT:______________
1(c). Drums Fiber (Plastic) DOT:______________
1(d). Weight (per drum):_____lbs. or Volume:____gal.

2. Bulk (Y,N):____Type/Size:______________________

3. Other (Y,N):___ (Describe):______________________

4. Transportation Requirements (Describe):______________________

5. Customer to Transport (Y,N):______________________
6. Truck Scale at Customer Site (Y,N)______________________

PAGE 8 HWMF PROFILE NO.:

M. POLYCHLORINATED BIPHENYL (PCB) ITEMS

1. LIQUIDS:
   a. Name of Original Fluid:________________________________
   b. Name of Solvent:____________________________________
   c. PCB Concentration: ________ppm
   d. Shipping Weight:___________kilograms

2. CAPACITORS:
   a. Quantity_________________
   b. Shipping Weight:___________

3. TRANSFORMERS AND REGULATORS:
   a. Dimensions:______________
   b. Nameplate Gallons:_________
   c. PCB Concentration: _________
   d. Shipping Weight:___________
   e. To Be Shipped: Full:________Drained Only:_______
   Drained and Flushed:(Y,N)____

4. OTHER TYPES OF PCB ARTICLES (Be specific):
   a. Name:__________________________
   b. Quantity:_______________________
   c. Shipping Weight:_________________
   d. PCB Concentration:_______________

N. RECOMMENDED TREATMENT METHOD (Check one 'X')

1. Incineration:________

2. Incineration with chemical stabilization and encapsulation of residues:_____

3. Chemical stabilization and encapsulation:______

4. Direct landfill:______________ 5. Underground injection:____________

6. Chemical treatment:_________ 7. Fuel substitution:_________________

8. Other:______(Describe):_____________________________________

O. CERTIFICATION

I hereby certify and warrant that the information supplied on this form and on any attachments or supplements represents a complete and accurate identification and description by the generator company of this waste material, its constituents and its known or suspected hazards.

Signature:____________________________ Date: _______/______/

Print Name:____

LINE-BY-LINE INSTRUCTIONS FOR COMPLETING SAMPLE PROFILE FORM

General Instructions

- The Waste Profile Form is for use by the on-site permitted storage facility, hereafter referred to as HWMF.
- Please type or print in black ink.
- Detailed instructions are included to help you complete the Waste Profile Form. The letters and numbers which precede each instruction refer to the lettered and numbered entries on the Waste Data Sheet.
- The Waste Profile Form must be signed by an authorized employee of the generator, as it is a contractual document.
- If you have any questions concerning the use of the Waste Profile Form, please contact the HWMF administrator at (__)________
- Make a copy of the completed Waste Profile Form for your records. Mail the ORIGINAL and all attachments to the HWMF administrator.

This information is used to determine if the waste may be treated, stored, or disposed in a legal, safe, and environmentally sound manner. This information will be maintained in strict confidence. ANSWERS MUST BE MADE TO ALL QUESTIONS. A response of "NONE" or "N.A." (not applicable) can be made if appropriate.

Itemized Instructions

SECTION A. GENERATOR INFORMATION

1. Generator Facility: Enter the name of the generating facility.
2. Technical Contact: Enter the name and title of the person.
3. Phone Number: Enter Technical Contact's phone number.
4. Emergency Contact, Agency: Enter the name of the agency which should be contacted in event of any emergency situation involving the waste.
5. Emergency Contact, Person: Enter the name and title of the person who should be contacted in the event of any emergency situation involving the waste.
6. Phone Number: Enter Emergency Contact Person's phone number.
7. Generator US EPA ID: Enter the ID issued by the US EPA to the facility generating the waste (if applicable).
8. Generator ID Number: Enter generator's corporate division code.
9. Generator Cost Code: Enter generator's cost code.

SECTION B. PICKUP LOCATION

1. Company Name: Enter the name of the company at the physical pickup location.

2. Group or Division: Enter group or division.
3-6. Address: Enter the address of the Company at the pickup location (not a P.O. Box).
7. Pickup Contact: Enter the name and title of the person who may be contacted at the pickup location.
8. Phone Number: Enter the Pickup Contact's phone number.

SECTION C. GENERAL WASTE DESCRIPTION

1. Name of Waste: Enter the name that is generally descriptive of the waste (e.g., Paint Stripper).
2. Process Generating Waste: List the specific process generating the waste.
3-4. Quantity Generated: Enter actual or estimated quantity to be disposed per period of time (year or month).

SECTION D. WASTE COMPOSITION

List all organic and/or inorganic compounds of the waste, using specific chemical name and constituent. DO NOT list trade names. For all components, indicate the concentration in which EACH component is present. The concentration may be expressed as a range (e.g., +/- 10%). The total of all concentrations must equal a minimum of 100%. For each component list the Chemical Abstract Substance Registry Number (CASRN) and the Reportable Quantity (RQ) in pounds.

SECTION E. PROPERTIES OF WASTE

1. Indicate color of waste (e.g., brown, clear, etc.)
2. DO NOT SMELL THE WASTE. Indicate if the waste has a known incidental odor (e.g., sweet, pungent, etc.)
3. Indicate the flash point in degrees Fahrenheit obtained per SW-846 as referenced in 40 CFR § 261.21.
4. For aqueous liquids indicate pH; for all other liquids and sludges indicate corrosivity (in mm/year), if known, per 40 CFR § 261.22.
5. Specify the melting point for a solid in degrees Fahrenheit.
6. Specify the boiling point for a liquid in degrees Fahrenheit.
7. Indicate specific gravity or bulk density in pounds per cubic foot.
8. Specify viscosity in centipoise at 25 degrees Celsius (or 77 degrees F) for liquids and sludges only.
9. Specify vapor pressure in mm of mercury and fill in appropriate temperature in degrees Celsius for liquids and sludges only.
10. Indicate Heat of Combustion in BTU/pound.
11. Specify concentration (in weight %) of Bromine.
12. Specify concentration (in weight %) of Iodine.
13. Specify concentration (in weight %) of Fluorine.
14. Specify concentration (in weight %) of Chlorine.

15. Specify concentration (in weight %) of Sulfur.
16. Indicate qualitative assessment of flowability of the waste at 25 degrees Celsius (or 77 degrees F).
17. Check all applicable boxes identifying physical state, by volume, at 25 degrees Celsius (or 77 degrees F). Indicate approximate % of each phase or layer for each physical state.
18. Indicate percent solid by weight.
19. Indicate Total Organic Carbon in mg/L.
20. Indicate the concentration of Total Volatile Organic Compounds (VOC) in mg/L per EPA method 8015, or 8010 and 8020.
21. Specify concentration (in weight %) of Halogenated Organic Compounds in mg/kg or mg/L.
22. Specify concentration of Halogenated Organic Compounds in mg/kg or mg/L.
23. Cyanides: Enter mg/kg or mg/L.

SECTION F. ELEMENTAL ANALYSIS

Total Metals and TCLP metals data are required for incinerable waste streams. TCLP metals data are required for landfill materials. For liquid wastes with less than 0.5% solids, total constituents (in mg/kg) x density of waste (in kg/L) = TCLP results (in mg/L). Record data for all that apply.

SECTION G. ORGANIC TOXIC CHARACTERISTIC WASTE

Specify the total analysis and/or TCLP results in mg/L for all that apply. For liquid wastes with less than 0.5% solids, total constituents (in mg/L) x density of waste (in kg/L) = TCLP results (in mg/L). Record data for all that apply.

SECTION H. REACTIVE CHARACTERISTICS

1-5. Indicate reactive characteristics of the waste by marking appropriate boxes to specify if waste is explosive, pyrophoric, shock sensitive, water reactive or air reactive.
6. Indicate if the waste contains reactive cyanide and specify concentration (in mg/kg).
7. Indicate if the waste contains reactive sulfide and specify concentration (in mg/kg).
8. Indicate if the waste contains oxidizers and specify.
9. Indicate if the waste has any other reactive characteristics and specify.

SECTION I. REGULATORY COMPLIANCE

1. Indicate and list OSHA carcinogens per list below and indicate ppm.
   Note: container labels must identify compounds and cancer warning.

   (1) Acrylonitrile (vinyl cyanide)
   (2) 2-Acetylaminofluorene
   (3) Aminodiphenyl
   (4) Asbestos
   (5) Benzidine
   (6) Benzene
   (7) bis-Chloromethyl Ether
   (8) 3,3'-Dichlorobenzidine (and salt)
   (9) Dimethylaminoazobenzene
   (10) 1,2-Dibromo-3-Chloropropane
   (11) Ethylenimine
   (12) Ethylene Oxide
   (13) Formaldehyde
   (14) Methyl Chloromethyl Ether
   (15) 4,4'-Methylene bis (2-Chloroaniline)
   (16) alpha-Napthylamine
   (17) beta-Napthylamine
   (18) 4-Nitro Biphenyl
   (19) N-Nitrosodimethylamine
   (20) beta-Propiolactone
   (21) Vinyl Chloride

2. Indicate if the waste contains radioactive material.
3. Indicate if the waste contains TSCA regulated PCBs.
   Indicate concentration range (ppm) and complete Section P:PCB Items. PCBs are TSCA regulated when PCB concentration of original PCB Source or PCB Waste is >50 ppm.
4. Indicate if the waste contains biomedical waste materials and/or infectious waste materials. (Note that a special manifest is required for disposal of biomedical and infectious waste in the state of New Jersey).
5. Indicate if the waste contains FIFRA pesticides with specific disposal requirements.
6. Indicate if the waste contains asbestos.
7. Indicate if the waste is a California List Regulated Waste per 40 CFR § 268.32.
8. Indicate if the waste is a Solvent Regulated Waste per 40 CFR § 268.30.
9. Indicate if the waste is a USEPA Dioxin Regulated Waste per 40 CFR § 268.31.

SECTION J. SHIPPING INFORMATION

1. Indicate if the waste is an EPA RCRA Hazardous Waste per 40 CFR Part 261.

2. Indicate if the waste is wastewater or non-wastewater. A wastewater is <1% Total Organic Compounds and <1% Total Suspended Solids.
3. List all applicable EPA Hazardous Waste Codes and subcategories (per 40 CFR Part 268) including characteristic ("D") codes per 40 CFR Part 261.
4. Indicate if the waste is a US DOT Regulated Material per 49 CFR § 172.101.
5. Indicate the proper US DOT shipping name for the waste per 49 CFR § 172.101.
6. Indicate other major constituents of the waste for DOT shipping name.
7. Enter the proper US DOT Hazard Class per 49 CFR § 172.101.
8. Enter the proper US DOT (UN/NA) identification number per 49 CFR § 172.101.
9. Enter the packaging group number per 49 CFR § 172.101.
10. Indicate proper US DOT Guide Number per DOT Emergency Response Guide Book.
11. Enter the reportable quantity for the waste (in pounds) per 49 CFR § 172.101 or 40 CFR Part 302 (if applicable).
12. Indicate the constituent for which the RQ was determined.
13. Indicate EPA Compatibility Group per 40 CFR Part 264, Appendix V.

SECTION K. MANIFEST INFORMATION/BILL OF LADING.

1. Enter the name of the company to which manifest questions or comments should be directed.

2-5.Enter the address of the company to which manifests should be returned.

6. Enter the name and title of the person to be contacted in the event that manifest questions or comments occur.
7. Enter the manifest contact's phone number.

SECTION L. METHOD OF SHIPMENT

1. Indicate type and size (e.g., DOT 17E/55 gallon; 2-5 gallon fiber drums in 55 gallon steel).
2. If waste is to be shipped in bulk, indicate type and size of equipment (e.g., 5000 gallon tank truck; 6000 gallon heated tank truck; 20 cu. yd. bin).
3. Indicate other methods of shipment (e.g., box, pallets).
4. Indicate any special transportation requirements (e.g., 40 feet of 2-inch hose; security clearance required prior to entry).
5. Indicate if customer is to handle transportation.
6. Indicate if truck scale available at customer site.

SECTION M. POLYCHLORINATED BIPHENYL (PCB) ITEMS:

1. Liquids
   a. Indicate name of original fluid (e.g., Transformer Oil, Askarel).
   b. Indicate name of any solvent contaminated with PCBs (e.g., diesel).
   c. Indicate PCB concentration (in ppm).
   d. Indicate shipping weight in kilograms.
2. Capacitors

   a. Indicate quantity of PCB capacitors.
   b. Indicate shipping weight of PCB capacitors in kilograms.
3. Transformers and Regulators.
   a. Indicate dimensions of transformers or regulators.
   b. Specify nameplate gallons.
   c. Indicate PCB concentration of transformers or regulators (in ppm).
   d. Specify shipping weight in kilograms.
   e. Indicate if transformers or regulators will be shipped full or drained only or drained and flushed.
4. Other types of PCB articles (be specific).
   a. Specify name of any other PCB articles (e.g., spill debris).
   b. Indicate quantity and shipping container.
   c. Indicate shipping weight in kilograms.
   d. Indicate PCB concentration in ppm.

SECTION N. RECOMMENDED TREATMENT METHOD

1-7.Check applicable box for recommended treatment method.
8. If treatment method other than listed in 1-6, indicate method.

SECTION O. CERTIFICATION

The Waste Data Sheet is a contractual representation of the waste material. The Waste Data Sheet must be completed and signed by a knowledgeable and authorized employee of the generator of the waste.

Analysis Request-- Determination of RCRA Characteristic

Sample No. ________________ Waste Stream Name______________
Date Sampled ______________ Waste Stream Profile No.__________
Drum No. _________________Generator______________________

INSTRUCTIONS: Mark the type of analysis you wish to have performed. If the analysis is not applicable, write N/A.

IGNITABILITY (D001):

Flash Point:________(Closed Cup)

Flash Point:________(Open Cup)

CORROSIVITY (D002):

pH Measurement:____________________
Corrosivity toward steel:_______________
Method SW 846 - 1110

REACTIVITY (D003):

Total and available Cyanide:____________
Method SW 846 - 9010 or 9012

Sulfides:___________________________
Method SW 846 - 9030

Reactivity with air:____________________

Reactivity with water:_________________

Reactivity with 10% by wt. solution HCL at a 1:1 ratio:_________

Reactivity with 10% by wt. solution NaOH at a 1:1 ratio:________

Reactivities per 40 CFR § 261.23_________

Analysis Request-- Characteristic Toxic Metals

Sample No.____________________Waste Stream Name____________
Date Sampled__________________Waste Stream Profile No._________
Drum No._____________________Generator____________________

INSTRUCTIONS: Mark the type of analysis you wish to have performed. If the analysis is not applicable write N/A.

NOTE: If the total analysis concentration for a contaminant is less than the regulated concentration the TCLP extraction does not have to be performed. However if the total analysis concentration for the contaminant is greater than 20 times, *for solids*, the regulated concentration , the TCLP procedure does have to be performed.

EPA NO.	CONTAMINANT	REG. CONC. (MG/L)	TOTAL ANALYSIS	TCLP ANALYSIS
D004	ARSENIC	5.0		
D005	BARIUM	100.0		
D006	CADMIUM	1.0		
D007	CHROMIUM	5.0		
D008	LEAD	5.0		
D009	MERCURY	0.2		
D010	SELENIUM	1.0		
D011	SILVER	5.0		

Analysis Request-- Characteristic Toxic Organics

Sample No. ________________ Waste Stream Name______________
Date Sampled ________________ Waste Stream Profile No.__________
Drum No. ___________________ Generator_______________________

INSTRUCTIONS: Mark the type of analysis you wish to have performed. If the analysis is not applicable write N/A.

NOTE: If the total analysis concentration for a contaminant is less than the regulated concentration the TCLP extraction does not have to be performed. However, for physically solid wastes, if the total analysis concentration for the contaminant is greater than 20 times the regulated concentration , the TCLP procedure may have to be performed.

EPA NO.	CONTAMINANT	REGULATED CONC.(MG/L)	TOTAL ANALYSIS	TCLP ANALYSIS
:D012	ENDRIN	0.02		
:D013	LINDANE	0.4		
:D014	METHOXYCHLOR	10.0		
:D015	TOXAPHENE	1.0		
:D016	2,4-D	10.0		
:D017	2,4,5-TP (SILVEX)	1.0		
:D018	BENZENE	0.5		
:D019	CARBON TETRACHLORIDE	0.5		
:D020	CHLORDANE	0.03		
:D021	CHLOROBENZENE	100.0		
:D022	CHLOROFORM	6.0		
:D023	o-CRESOL	200.0		
:D024	m-CRESOL	200.0		
:D025	p-CRESOL	200.0		
:D026	CRESOL	200.0		
:D027	1,4-DICHLOROBENZENE	7.5		
:D028	1,2-DICHLOROETHANE	0.5		
:D029	1.1-DICHLOROETHYLENE	0.7		
:D030	2,4-DINITROTOLUENE	0.13		
:D031	HEPTACHLOR (& its hydroxides)	0.008		
:D032	HEXACHLOROBENZENE	0.13		
:D033	HEXACHLOROBUTADIENE	0.5		
:D034	HEXACHLOROETHANE	3.0		
:D035	CARBON TETRACHLORIDE	0.5		
:D036	METHYL ETHYL KETONE	200.0		
:D037	PENTACHLOROPHENOL	100.0		
:D038	PYRIDINE	5.0		
:D039	TETRACHLOROETHYLENE	0.7		
:D040	TRICHLOROETHYLENE	0.5		
:D041	2,4,5-TRICHLOROPHENOL	400.0		
:D042	2,4,6-TRICHLOROPHENAL	2.0		
:D043	VINYL CHLORIDE	0.2		

Analytical Request-- Determination of F001 - F005 Solvents

Sample No.____________________Waste Stream Name____________
Date Sampled__________________Waste Stream Profile No.________
Drum No.______________________Generator____________________

INSTRUCTIONS: Mark the type of analysis you wish to have performed. If the analysis is not applicable write N/A.

F001 AND F002:

Tetrachloroethylene:______ Chlorinated Fluorocarbons:_______

Trichloroethylene:________ Carbon Tetrachloride:___________

Methylene Chloride:______ 1,1,1 Trichloroethane (TCA):_____

F002:

Chlorobenzene:__________ Trichlorofluoromethane:_________

Otho-dichlorobenzene:_____ 1,1,2-Trichloroethane:___________

F003:

Xylene:_________________ Acetone:_____________________

Ethyl Benzene:___________ n-Butyl Alcohol:_______________

Ethyl Ether:_____________ Cyclohexanone:________________

Methanol:_______________

F004

Cresols:_________________ Cresylic Acid:_________________

Nitrobenzene:______

F005

Toluene:________________ Methyl Ethyl Ketone:_________________

Carbon Disulfide:________ Isobutanol:____________________

Pyridine:________________ Benzene:______________________

2-Ethoxyethanol:_________ 2-Nitropropane:________________

CHAPTER 7

STORAGE OF HAZARDOUS WASTE

There are four types of allowed hazardous waste storage facilities: satellite accumulation areas, 90-day accumulation areas, permitted storage facilities, and 10-day transfer facilities. This chapter concerns itself with the first three.

Satellite accumulation areas are those areas where waste is first accumulated by the generator. The areas should be conveniently located near the point of generation so that as waste is generated it may be placed in secure containers in a safe, delineated location.

90-Day accumulation areas are facilities that accept waste from the satellite accumulation areas for bulking and transfer of the waste either to the ultimate treatment, storage, and/or disposal facility or to a permitted generator storage facility (hereafter referred to as the permitted hazardous waste management facility or "HWMF").

The satellite accumulation areas and 90-day accumulation areas are *not* permitted facilities. For the generator to store waste after 90 days (may be longer depending on distance from disposal facility and generator size) he must have a permitted facility. The next chapter will provide information for obtaining a Part B permit for a HWMF.

Permitted transporters of hazardous waste may store waste up to 10 days at transfer stations.

SATELLITE ACCUMULATION AREAS

Separate satellite areas may be established for each type of waste stream where separating the wastes allows the generator to handle them in a safer and more effective manner, and to segregate the wastes for proper disposal. For example, three 55-gallon drums of three different types of hazardous waste may exist in the same satellite collection area provided that they are identified in the facility's contingency plan and otherwise meet the requirements for satellite areas.

40 CFR § 262.34(c)(1) describes accumulation limits near any point of generation. Since the limit is on the point of generation rather than the facility, the satellite accumulation regulations allow up to one quart of acutely hazardous waste or 55 gallons of hazardous waste for ***an identified hazardous waste stream in each satellite area***. Using a small container to transfer contents to a larger container at or near the point of generation is an acceptable practice provided the maximum volume (one quart of acutely hazardous waste and 55 gallons of hazardous waste) for that particular waste stream is not exceeded and both containers are at or near the point of generation.

Additional Requirements for Satellite Areas

Equipment must be in sufficient quantities and appropriate locations to be able to respond to fires, spills, or other emergency situations at each satellite area at the facility.

Compliance with fire codes and OSHA requirements would satisfy certain state Hazardous Waste Management (CHWM) regulations pertaining to placement of portable fire extinguishers to the extent that this equipment is needed at a facility. However, portable fire extinguishers are not required at each satellite accumulation area if other suitable fire control equipment (e.g., overhead fire extinguishing equipment) is available for each area.

The necessary required equipment at a given facility depends upon the type of hazards posed by the waste handled at the facility (see 40 CFR § 265.32). For example, the Colorado Department of Health interprets these regulations to mean each facility at a minimum must have: a system for internal emergency communications; a device for communication with external police, fire and emergency response agencies; and adequate fire control equipment. Spill control and decontamination equipment would be required at most facilities and would need to be designed specifically for the types of materials handled.

Good Practices for Satellite Accumulation Areas

Each satellite accumulation area (SAA) must be at or near the point of waste generation and under the control of the process operator.

Containers used for accumulation must be marked with a "Hazardous Waste" label (see chapter 12) as soon as the initial hazardous waste is added or as soon as the contents of the container are characterized as hazardous waste. This label, at a minimum, should identify the type of waste accumulated.

A maximum of 55 gallons of hazardous waste may be accumulated, whereas a maximum of only 1 quart of acutely hazardous waste may be accumulated for each waste stream.

Smaller containers may be used, provided the 55-gallon/1 quart waste limit is not exceeded.

Wastes placed in a given container must be compatible with each other. Wastes in separate containers should be grouped according to compatibility.

Containers must be chemically compatible with the wastes being accumulated. Containers must be kept closed except when adding, removing or inspecting waste. Also:

- A funnel must not be left in the container if waste is not being added or removed.
- Bottles, jugs, jars, etc., must have screw-type lids and they must be firmly attached.
- Bung-opening drums must have the bung closed.

Liquid waste containers should be provided with approximately two inches of freeboard to allow for expansion.

When containers are moved, it shall be done in such a manner as to minimize the opportunity for damage, ruptures and leaks. Containers being moved must be tightly sealed.

A Hazardous Waste Container Log may be used to record all wastes being accumulated.

When the satellite accumulation area volume limit is reached (i.e., 55 gallons hazardous waste or 1 quart acutely hazardous waste for each generation point), the date the limit is reached must be marked on the container label and the waste moved to:

- an approved 90-day accumulation area, or
- a permitted storage/treatment area.

Waste is not allowed to move from one satellite accumulation area to another satellite accumulation area. The acceptable flow of waste is (1) from a satellite accumulation area to a 90-day accumulation area, or (2) from a satellite accumulation area to a permitted storage area, or (3) from a satellite accumulation area to a permitted treatment facility.

Tracking of the waste volume is important to ensure that the 55-gallon (1 quart acutely hazardous waste) limit is not exceeded. Any container approaching the capacity limit should be identified so that preparation can be made for its proper transfer.

The hazardous wastes in the container must be assessed to determine if the wastes are subject to the Land Disposal Restrictions (LDRs). This must be noted on the label so that the one-year time limit for LDR wastes (for certain generators) can be tracked.

Care should be taken to ensure that hazardous waste is not mixed with non-hazardous waste because the entire amount, if mixed, would be classified as hazardous.

The satellite accumulation area must have a sign posted that reads, "SATELLITE ACCUMULATION AREA." "NO SMOKING" signs must be posted in areas with ignitable or reactive wastes. Adequate aisle space must be maintained. (A two-foot aisle space is recommended.)

90-DAY ACCUMULATION AREAS

90-day accumulation areas differ from satellite accumulation areas in that there is a fixed maximum accumulation time limit a container may be kept in storage (90 days) but no volume limit for the amount of waste stored. Most of the good practice rules that apply to satellite collection facilities also apply to 90-day storage facilities. Note the following practices as well.

- The 90-day accumulation area should have a non-permeable surface (e.g., concrete, plastic) and the ability to catch and direct spills for easy cleanup.
- There must be a contingency plan for spills and to delineate emergency procedures.
- The individual(s) in charge of the area should be trained and should keep appropriate records regarding inventory and time limits.
- 90-day accumulation areas are subject to closure requirements.
- The area should be inspected weekly.
- Ignitable and/or reactive waste must be stored more than 50 feet from the property line, and incompatible waste should be stored in separate locations.
- All containers in the accumulation area must be labeled and marked with accumulation start dates.
- There must be adequate aisle space for movement of containers and inspections; usually, drums are stacked only two high and two wide in a row.

PERMITTED STORAGE FACILITIES (HWMFs)

If a generator has numerous 90-day collection areas, he *may* find that a central permitted storage facility that can store waste for longer periods may be economically justifiable. Some facilities also conduct waste treatment in addition to bulking and staging for off-site disposal. Be aware that obtaining a Part B permit for the facility will be costly and time consuming. It is advisable that *all* options are explored before deciding on a permitted facility. The next chapter goes into detail concerning the permitting of such a facility.

CHAPTER 8

PERMITTING OF TREATMENT, STORAGE OR DISPOSAL FACILITIES (PART B PERMITS)

If a facility has already completed EPA form 8700-12, Notification of Regulated Waste Activity, then it has already completed Part A of a two-part application. Part A permitting (see Chapter 3) is a relatively straightforward process in which the applicant simply completes the form. Part B is a much more complex undertaking. A facility would complete a Part B permit if it wants to store waste for more than 90 days (180 days for small quantity generators), conduct treatment, or dispose of waste on-site.

Since the Part B permitting process is quite long, arduous and expensive, the RCRA administrator should look into all available options which would obviate the need for a Part B permit. If there is no escaping the need for a Part B permit, the following section will provide the basics of the permitting process.

OVERVIEW

Part B permits identify the administrative and technical standards to which facilities must adhere. Permits can be issued by EPA or an authorized state. Whether administered by EPA or a state, the permitting program must meet national standards. Indeed, one of the requirements for a state program is that it be consistent with, and no less stringent than, the federal program. Therefore, although this chapter describes permitting as a federal program, the procedures outlined apply equally to permitting programs run by authorized states. Additionally, states may impose regulatory requirements that are more stringent or broader in scope than the federal program.

This chapter covers the entire permitting process including:

- The universe of treatment, storage and disposal facilities subject to permitting requirements of Subtitle C of RCRA.
- The steps involved in permitting a Treatment Storage and/or Disposal Facility (TSDF).
- Corrective action through the permitting process.

WHO NEEDS A PART B PERMIT?

Owners or operators of facilities that treat, store, or dispose of hazardous waste must obtain an operating permit under Subtitle C of RCRA. This definition covers large quantity generators who want to store waste at their facility for longer than 90 days. New TSDFs must receive a RCRA permit before construction can commence. There are, however, some exclusions to this requirement. These include:

- Generators storing waste on site for less than 90 days;
- Farmers disposing of their own (hazardous) pesticides on site;
- Small quantity generators who store waste on site less than 180 days(270 days in some cases);
- Owners or operators of totally enclosed treatment facilities, wastewater treatment units (tanks) and elemental neutralization units;
- Transporters storing manifested wastes at a transfer facility for less than 10 days;
- Persons engaged in containment activities during an immediate response to an emergency;
- Persons adding absorbents to waste as long as it coincides with placement of waste in the container; and
- Owners or operators of solid waste disposal facilities handling only conditionally exempt small quantity generator waste. (Please refer to state regulations.)

If any of the individuals listed above treat, store, or dispose of hazardous waste in a manner not covered by the exclusion, they are subject to the RCRA permit requirements for that activity.

As noted earlier, a permit defines a facility's requirements under Subtitle C. These requirements consist of all the general and technical standards applicable to TSDFs, as well as requirements for corrective action. Corrective action requires all TSDFs to clean up releases caused by facility operations.

SPECIAL SITUATIONS

Included among the ranks of facility owners or operators required to obtain a permit under a RCRA Subtitle C are groups eligible for unique permits. These groups include owners or operators that:

- Have a permit under certain other environmental laws, or
- Have just constructed an incinerator or a land treatment facility.

Other Environmental Laws

EPA issues permits under a number of different laws. In some instances, the requirements of one statute's permitting regulations are quite similar to those in another statute. To avoid duplication, EPA has tried to abbreviate the application process for facilities that need to be permitted under two or more statutes. This is done through a permit-by-rule. A permit-by-rule eliminates the need for facilities to submit a full Subtitle C permit application when they are permitted under the:

- Safe Drinking Water Act (Underground Injection Control permit);
- Clean Water Act (National Pollutant Discharge Elimination System permit);or
- Marine Protection, Research, and Sanctuaries Act (Ocean Dumping permit).

Facilities seeking a RCRA permit that already have one of the three permits listed above need only meet a subset of the Subtitle C regulatory requirements. For example, an owner or operator of a barge or vessel that has an ocean dumping permit, and complies with the appropriate conditions under Subtitle C (e.g., obtaining an EPA ID number, using the manifest system, and biennial reporting) will be considered to have a permit under RCRA.

New Incinerator or Land Treatment Facility

EPA issues permits to construct and operate new hazardous waste management facilities. Such facilities cannot be constructed until a permit is issued. There is, however, an exception to this rule. Land treatment facilities and incinerators must go through a trial period during which their ability to perform properly under operating conditions is tested. This period is called a trial burn for incinerators and a land treatment demonstration for land treatment facilities.

Owners or operators of these two types of facilities are required to obtain temporary permits that are enforced while the facility is being tested. Once the facility adequately completes its test, the owner or operator can apply to modify the permit. This sets the final operating conditions based on the data generated from these demonstrations.

TYPES OF PERMITS

Four categories of permits are issued under the RCRA Subtitle C program. Each category defines operating requirements and various provisions specific to the permitting need.

- Treatment, Storage, and Disposal Permits
  Most commonly, RCRA permits are issued for treatment, storage, and disposal units. The units are: containers, tank systems, surface impoundments, waste piles, land treatment units, landfills, incinerators, and miscellaneous units. These methods are the most common way to treat, store, and dispose of hazardous waste. Minimum national standards have been promulgated for each of these methods (refer to 40 CFR Part 264). As part of the permitting process, HSWA (Hazardous and Solid Waste Amendments of 1984) requires facilities to correct releases to all media. Facilities must develop a schedule of compliance to address releases from all solid waste management units, as described later in this chapter.

- Research, Development, and Demonstration Permits - EPA encourages the use of alternate treatment technologies by issuing research, development, and demonstration (RD&D) permits for promising innovative and experimental treatment technologies.

- Post-Closure Permits - Land disposal facilities that do not close "clean" must obtain a post-closure permit, specifying the requirements for proper post-closure care.

- Emergency Permits - In potentially dangerous situations, EPA can forego the normal permitting process. Specifically, when there is an "imminent and substantial endangerment to human health and the environment," a temporary (90 days or less) emergency permit can be issued to a:

  - Non-permitted facility for the treatment, storage, or disposal of hazardous waste, or
  - Permitted facility for the treatment, storage, or disposal of hazardous waste not covered by its existing permit.

THE PERMIT PROCESS

All hazardous waste TSDFs required to get a RCRA permit go through the same basic permitting process. The exceptions are facilities that are issued a permit-by-rule or an emergency permit. The permit process consists of the following steps:

- Submitting a permit application;
- Reviewing the permit application;
- Preparing the draft permit;

- Obtaining exposure information and performing health assessments for surface impoundments or landfill facilities;
- Taking public comment and finalizing the permit; and
- Modifying and terminating a permit.

An additional step of appealing the permit decision may occur with some permits.

There are a number of federal laws that may affect the permit process, including the:

- Wild and Scenic Rivers Act;
- National Historic Preservation Act of 1966;
- Endangered Species Act;
- Coastal Zone Management Act; and
- Fish and Wildlife Coordination Act.

When any of these laws is applicable, its procedures must be followed. For example, the Coastal Zone Management Act prohibits EPA from issuing a permit for an activity affecting land or water use in the coastal zone unless the proposed activity complies with the state's Coastal Zone Management Program, and is agreed to by the state. For more information on these laws and their potential impacts on Subtitle C's permitting process, see 40 CFR § 270.3.

The Permit Application

Owners or operators of facilities that fall under the permitting regulations are required to submit a comprehensive permit application covering all aspects of the design, operation, and maintenance of the facility. This gives EPA and the state valuable information to determine if the facility is in compliance with Subtitle C regulations and to develop a facility-specific permit.

The permit application is divided into two parts: A and B. Part A (form 8700-12) of the application is a short, standard form that collects general information about a facility, e.g., name of the applicant and a description of the activities conducted at the facility (see Chapter 4). Part B of the permit application is much more extensive than Part A. It requires the owner or operator to supply detailed and highly technical information; e.g., chemical and physical analyses of the hazardous waste to be handled at the facility. Since there is no standard form for Part B, the owner or operator must rely on the regulations (40 CFR Parts 264 and 270) to determine what to include in this part of the application (see sample permit outline, Chapter 9). In addition to the general Part B information that must be submitted by all owners or operators of TSDFs, there are unique information requirements that are tied to the type of facility seeking a permit.

Depending on the situation, Parts A and B may be submitted at different times. Existing facilities (i.e., those that received hazardous waste on or before November 19, 1980), submitted their Part As when applying for interim status. Facilities that choose to extend their storage time limit or add treatment capabilities will submit a Part B. Facilities that have interim status may have their Part B permit voluntarily submitted or "called in" by EPA.

Under HSWA another group of facilities can submit Parts A and B separately. Specifically, any TSDF that comes under the jurisdiction of Subtitle C due to statutory or regulatory changes must submit its Part A six months after the date of publication of the regulations in the Federal Register, or 30 days after the date they first become subject to the promulgated standards. The Part B for such facilities can either be voluntarily submitted or called in by EPA. A special timetable applies to land disposal facilities that come under the jurisdiction of Subtitle C in this manner; namely, they must apply for a Part B within 12 months of becoming subject to Subtitle C requirements or lose interim status. Incinerators and all other facilities will retain interim status until a final permit determination is made if they submit their Part B applications by the indicated deadlines.

New facilities (ones commencing operations or construction after November 8, 1984), because they are ineligible for interim status, submit Part A and B simultaneously. This submission must be made at least 180 days prior to the date on which physical construction is expected to start. By requiring submittal of Parts A and B prior to construction, the owner or operator does not risk losing a substantial financial investment by building a facility that fails to meet the more stringent permit (40 CFR Part 264) requirements.

Reviewing the Permit Application

Once the owner or operator of a facility has submitted an application (both Parts A and B), EPA's first step is to determine if all the required information has been submitted. If the application is not complete, a notice of deficiency (NOD) letter is sent to the owner or operator describing the additional information that is required for a complete application. Once the owner or operator has submitted all of the required information, the application is considered complete. Failure to provide this information can result in denial of the permit, enforcement action, or both.

In some cases information contained in the permit application may be considered confidential by the owner or operator. Permit applicants often make a claim of confidentiality to protect trade secrets. In such cases, the owner or operator must make the claim known at the time of submission by following the procedures described in 40 CFR § 270.12 ("confidentiality of information"). Claims of confidentiality are reviewed (by EPA's legal counsel) to determine if the information can legitimately be claimed as confidential. If a claim is substantiated, the information is treated as confidential and not released. If, on the other hand, a claim is denied, the information is made public.

Once the owner or operator is informed, by letter, that his application is complete, an in-depth evaluation of the permit application begins. The purpose of the evaluation is to determine if the application satisfies the technical requirements of RCRA. After the permit application is evaluated, EPA makes a tentative decision either to issue or deny the permit. If the tentative decision is to deny the permit, EPA must send the owner or operator a notice of intent to deny. If EPA tentatively decides to issue the permit, a draft permit is prepared by EPA staff.

For new facilities that submit their applications after November 8, 1984, HSWA places no time limits on how long EPA can take to evaluate the application. In either case, evaluating an application is a lengthy process, and can take from one to three years.

Preparing the Draft Permit

The draft permit incorporates applicable technical requirements and other conditions pertaining to the facility's operation. These other conditions are divided into two groups - those applicable to all permits and those applied on a case-by-case basis. General permit conditions comprise:

- A requirement to comply with all conditions listed in the permit;
- A responsibility to notify EPA of any planned alterations or additions to the facility;
- A requirement to provide EPA with any relevant information requested and to allow Agency representatives to inspect the facility premises under certain conditions;
- A requirement to certify annually that a program is in place to reduce the volume and toxicity of waste, and that the proposed method of treatment, storage, and disposal minimizes threats to human health and the environment; and
- A duty to submit required reports, e.g., Unmanifested Waste Report, Biennial Report, and Manifest Discrepancy Report..

The case-by-case permit conditions include:

<u>Compliance Schedules</u> - These schedules are allowable only to bring a facility into compliance with corrective action requirements; all other permit conditions must be met prior to issuance.

<u>Duration of Permit</u> - The permit is valid for up to ten years; land disposal permits must be reviewed every five years.

Exposure Information and Health Assessments

HSWA Section 3019 requires that final permit applications for surface impoundments and landfills be accompanied by information on the potential for public exposure to hazardous wastes or constituents from facility releases. Congress' rationale is that landfills and surface impoundments may pose a greater health risk than other types of disposal facilities.

Once the exposure information is submitted, EPA makes it available to the Agency for Toxic Substances and Disease Registry (ATSDR). If EPA believes that the release poses a substantial risk to human health, the Agency requests that ATSDR perform a health assessment. The exposure information must at least address:

- Reasonably foreseeable potential releases from both normal operations and accidents at the facility, including releases associated with transportation to or from the facility;
- The potential pathways of human exposure to hazardous wastes or constituents resulting from the releases described above; and
- The potential magnitude and nature of the human exposure resulting from the releases described above.

Taking Public Comment and Finalizing the Permit

Once the draft permit (or notice of intent to deny) is completed, EPA is required to give public notice and allow 45 days for written comment. In certain cases a public hearing may also be held during this time. Along with the public notice EPA must issue either a fact sheet or a statement of basis to inform concerned parties about the permitting process that is taking place. Fact sheets are prepared for every major facility and any facility subject to widespread public interest, as determined by EPA. They include detailed information pertaining to the nature of the facility, the contents of the draft permit (or notice of intent to deny), and the procedures to be used in reaching the final administrative decision on the permit application.

In lieu of a fact sheet, a statement of basis may accompany a draft permit (or notice of intent to deny). A statement of basis is essentially a summarized version of the fact sheet. These supporting documents are sent to the applicant and, on request, to any other interested person.

If information submitted during the initial comment period appears to raise substantial new questions concerning the permit, the Agency must re-open or extend the comment period. In this situation the Agency may also decide to revise the draft permit (or notice of intent to deny).

After the comment period closes, the Administrator makes the final permit decision: the permit is either issued or denied. This decision may be appealed to the Administrator. When administrative appeals are exhausted, the petitioner may seek judicial review of the final permit decision.

Modifying and Terminating the Permit

Once issued, RCRA permits are valid for up to ten years. Land disposal permits have an additional requirement of being reviewed after five years. During the term of a permit, situations may arise which may cause the permit to be *Modified, Revoked and Reissued, or Terminated.*

Permit Modification

Permits may need modification for a number of reasons, including:

- Substantial alterations or additions to the facility;
- New information about the facility becomes available; or
- New statutory or regulatory requirements affect existing permitted activities.

In September 1988, EPA published regulations (under 40 §§ CFR 270.41 and 270.42) that revised permit modification procedures for changes that facility owners and operators may want to make. EPA categorized selected permit modifications into three classes and established administrative procedures for approving modifications in each class. It is important to note that the "major/minor" modification requirements utilized prior to September 1988 may still be implemented by states that chose not to adopt the new provisions.

The permit modification regulations provide owners and operators more flexibility to change permit conditions, expand public notification and participation opportunities, and allow for expedited approval if no public concern exists for a proposed modification.

The permit modification classes are defined as:

Class 1: Routine changes and correction of errors.

Class 2: Common or frequently occurring changes needed to maintain a facility's capability to manage wastes safely or conform to new requirements.

Class 3: Major changes that substantially alter the facility or its operations.

In addition to establishing permit modification classes and administrative procedures, this regulation also gives EPA the authority to grant temporary authorization for facilities to respond promptly to changing conditions.

Revocation and Reissuance of the Permit

EPA may revoke and reissue a permit in two circumstances:

- When cause exists for terminating the permit (under the circumstances described below), but EPA decides that revocation and reissuance is a more appropriate step; or
- When the permit holder plans to transfer the permit.

Permit Termination

In some instances, operators may not comply with the requirements stipulated in the permit, even after enforcement action. In this case it may be necessary to terminate a hazardous waste permit. EPA may terminate a permit or deny its renewal for three reasons:

- Noncompliance by the permittee with any condition of the permit;
- Failure on the part of the permittee to disclose any relevant information during the permit process or misrepresentation of facts at any time; or
- The permitted activity endangers human health and the environment and can only be regulated to acceptable levels by permit termination.

A facility whose permit is terminated must implement its closure plan as required under 40 CFR Part 264 Subpart G, Closure and Post-Closure.

SUMMARY

Permits detail the administrative and technical performance standards that TSDFs must adhere to, and thus are the key to implementing Subtitle C regulations. Owners and operators of existing or new facilities must (with a few exceptions) obtain an operating permit. Each TSDF permit must include provisions for corrective action to address releases from solid waste management units if a release has been detected. Special permit requirements pertain to permit-by-rule facilities, and facilities demonstrating the efficacy of their treatment technology (trial burns and land treatment demonstrations).

The permitting process has five steps:

1. Submitting a permit application
2. Reviewing the permit application
3. Preparing the draft permit
4. Taking public comment and finalizing the permit
5. Modifying and terminating a permit.

The final decision on the permit also may be reviewed by the EPA Administrator and appealed to the local U.S. District Court.

HSWA has greatly expanded cleanup requirements at RCRA facilities. Through a process called corrective action, facilities must remedy releases threatening human health and the environment. Corrective action has four main parts:

1. RCRA facility assessment
2. RCRA facility investigation
3. Corrective measure study
4. Corrective measure implementation.

Although usually done through the permitting process, corrective action may also be completed through an enforcement order.

CHAPTER 9

SAMPLE HAZARDOUS WASTE MANAGEMENT FACILITY(PART B) PERMIT OUTLINES

The Part B Permit is unique to each Treatment, Storage, and Disposal Facility (TSDF). The guidelines for the creation of the Part B application were presented in Chapter 5. This abbreviated chapter presents tables of contents from two TSDF Part B Permits that are specifically for a storage facility that conducts limited treatment. We hope the presentation of the permit outline will give the reader an appreciation of the necessary content of a Part B Permit.

Permittee:

ID Number:

Permit Number:

Pursuant to the Colorado Hazardous Waste Act (Title 25, Article 15, Sections 101 *et seq.*, hereafte called "the Act," and regulations promulgated thereunder by the Colorado Board of Health (codifie in Title 6 of the Code of Colorado Regulations), a permit was issued to operate a hazardous wast management facility (HWMF) consisting of storage and treatment units located at ______________ ________________, also located at latitude ____ and longitude ______. This permit is bein modified to incorporate recent changes in operation, consistent with 6 CCR 1007-3 Parts 260 throug 267, 99 and 100 (April 1990).

This permit is effective as of September 1, 1985, and will remain in effect through August 31, 199 unless revoked and reissued or terminated pursuant to 6 CCR, Section 100.61.

MODIFIED: ______________________, 1990

SIGNED: ____
Date

John Doe
Unit Leader
Hazardous Waste Facilities
Hazardous Materials and Waste Management Division
Colorado Department of Health

EXAMPLE 1

TABLE OF CONTENTS

1.12.1 Job Titles and Duties

1.12.2 Training Content and Frequency

1.12.3 Training Records

1.13 TRAINING FOR EMERGENCY RESPONSE

1.14 GENERAL REQUIREMENTS FOR IGNITABLE, REACTIVE OR INCOMPATIBLE WASTE

1.15 PREPAREDNESS AND PREVENTION

1.15.1 Equipment

1.15.2 Testing and Maintenance of Equipment

1.15.3 Access to Communications or Alarm System

1.16 CONTINGENCY PLAN

1.17 MANIFEST SYSTEM

1.17.1 Use of Manifest

1.17.2 Completing the Manifest

1.17.2.1 Incoming Waste Shipments

1.17.2.2 Manifest Discrepancies

1.17.2.3 Outgoing Waste Shipments

1.17.3 Unmanifested Waste

1.18 LAND DISPOSAL RESTRICTIONS

1.19 RECORDKEEPING AND REPORTING

1.19.1 Operating Record

1.19.2 Waste Minimization

1.19.3 Annual Report

1.20 AVAILABILITY, RETENTION AND DISPOSITION OF RECORDS

1.21 CLOSURE

1.22 COST ESTIMATE FOR FACILITY CLOSURE

1.23 FINANCIAL ASSURANCE FOR FACILITY CLOSURE

1.24 LIABILITY SURETY

1.25 INCAPACITY OF OWNER OR OPERATORS, GUARANTORS OR FINANCIAL INSTITUTIONS

2.0 SPECIFIC CONDITIONS FOR STORAGE IN CONTAINERS

2.1 WASTE IDENTIFICATION

2.2 STORAGE CAPACITY

2.3 CONTAINER STORAGE PAD LAYOUT

2.4 AISLE SPACE

2.5 CONDITION OF CONTAINER

2.6 WASTE ANALYSIS

2.7 COMPATIBILITY OF WASTE WITH CONTAINERS

2.8 MANAGEMENT OF CONTAINERS

2.9 INSPECTION

2.10 CONTAINER STORAGE PAD DESIGN

2.11 CONTAINMENT

2.12 REMOVAL OF LIQUIDS FROM CONTAINMENT TANKS

2.13 SPECIAL REQUIREMENTS FOR IGNITABLE OR REACTIVE WASTE

2.14 SPECIAL REQUIREMENTS FOR INCOMPATIBLE WASTE

2.15 LAB PACKS

3.0 SPECIFIC CONDITIONS FOR BIOLOGICAL TREATMENT IN TANKS

3.1 WASTE IDENTIFICATION

3.2 TREATMENT FACILITY DESCRIPTION

3.3 FEED SYSTEM OPERATION

3.4 OVERFILL AND SPILL PREVENTION

3.5 SECONDARY CONTAINMENT

EXAMPLE 2

TABLE OF CONTENTS

PART III STANDARD CONDITIONS

A. Effect of Permit
B. Permit Actions
C. Severability
D. Duty and Requirements
   1. Duty to Comply
   2. Duty to Reapply
   3. Permit Expiration
   4. Need to Halt or Reduce Activity Not a Defense
   5. Duty to Mitigate
   6. Proper Operation and Maintenance
   7. Duty to Provide Information
   8. Inspection and Entry
   9. Monitoring and Records
   10. Reporting Planned Changes
   11. Anticipated Noncompliance
   12. Transfer of Permits
   13. Compliance Schedules
   14. 24-Hour Reporting
   15. Other Noncompliance
   16. Other Information
E. Signatory Requirement
F. Confidential Information
G. Documents to be Maintained at Facility Site

PART IV SPECIFIC CONDITIONS FOR BIOLOGICAL TREATMENT IN TANKS

A. Treatment Facility Description
B. Waste Identification
C. Secondary Containment Requirements
D. Operating Requirements
E. Response to Leaks or Spills
F. Inspection Schedules and Procedures
G. Recordkeeping and Reporting
H. Closure and Post Closure Care
I. Special Conditions for Ignitable or Reactive Wastes
J. Special Conditions for Incompatible Wastes
K. Waste Analysis

LIST OF ATTACHMENTS

Attachment A- Waste Analysis Plan
Attachment B- Inspection Schedules
Attachment C- Personnel Training
Attachment D- Contingency Plan
Attachment E- Closure Plan and Closure Cost Estimate
Attachment F- Container Storage Area Design
Attachment G- Digester Tank Treatment System

CHAPTER 10

PREPARATION OF A WASTE ANALYSIS PLAN

A Waste Analysis Plan (WAP) is required for all generators who possess a Part B Permit; all treatment, storage and disposal facilities. In Chapter 6 we delineated the first half of a waste analysis plan; that is, the initial profiling of waste streams. All generators should profile their waste streams. The second half of a waste analysis plan is confirming that the profiled waste conforms to the expected profile conditions prior to storage or disposal.

The method used to test individual shipments of waste is called a fingerprinting test or screen. The objective of the fingerprinting test is to document that the waste meets expected profile conditions such as pH, number of phases, reactivity, etc. Sampling procedures and fingerprinting test procedures are presented in this chapter.

SCOPE AND PURPOSE OF A WASTE ANALYSIS PLAN

The Waste Analysis Plan (WAP) is an analytical blueprint for specific waste analysis and quality control check for waste profiles. It initially includes profiling all waste streams. All waste streams must be analyzed unless process knowledge provides a very high degree of certainty as to the composition of the waste stream.

Some confusion exists regarding the necessity to analyze recycled hazardous waste. There is no regulatory requirement for such analysis as long as the composition can be determined via process knowledge.

REGULATORY REQUIREMENTS

Waste Analysis Plans are required for all generators who have a storage permit (40 CFR Part 264 & § 265.13(a)). They are not required if you are only storing waste for less than 90 days and *do not have a permit.* However, no reputable hazardous waste disposal firm will accept waste that has not been profiled.

Waste Analysis Plans are required for all Treatment Storage and Disposal Facilities (40 CFR § 268.7). Note: *state and EPA enforcement representatives have focused on the requirement to have a WAP since the passing of the land disposal restrictions, because the land disposal regulations are driven by waste composition and concentrations. In order to complete the Land Disposal Restrictions the composition of the waste must be determined.*

PROFILING AND FINGERPRINTING

The fingerprinting procedure to be presented is the second half of a WAP program. The first half is the waste profiling procedure.

ANALYSIS VERSUS USER'S KNOWLEDGE

Regulations allow the generator to use process knowledge, as well as process and raw materials inputs to determine the waste composition. We recommend generators perform a baseline analysis on most wastes during the profiling and rely on user's knowledge to determine if a significant change has taken place. As you will see, the fingerprinting process is also designed to catch changes and irregularities in waste streams. Often the disposal facility will require analysis as a condition of removal, making the discussion of process knowledge a moot point.

PREPARATION FOR SAMPLING

Drums to be sampled should be moved to a separate staging area prior to opening. The staging area would be clear of all other drums, equipment and traffic to prevent the spread of contamination and possible fire or explosion.

Personal Protective Equipment (PPE). The level of protection for drum sampling is determined by research on drum contents and by generator and laboratory results. If a lesser degree of respiratory protection is chosen (Level D or C), continuous air monitoring must occur during drum sampling activities to ensure that personnel are not being overexposed to volatile materials.

Drum Opening. Manual methods with non-sparking metal alloy tools are used when drums show no signs of over-pressurization, damage, or corrosion, and are not air-reactive or explosive. If drums show visual signs of pressurizing, corrosion, or certain reactives or explosives, then remote opening devices are recommended. Nondestructive techniques are to be used whenever possible to open drums for sampling.

General Sampling Procedures

Liquid samples from drums are usually collected with glass tubes. The glass tubes are normally 122 cm long and 6 to 8 mm inside diameter. The larger diameter tubes may be necessary to sample viscous liquids. The glass tubing is placed into the open drum. At least 30 cm of tubing should remain above the drum or liquid to prevent direct contact with the contaminated materials. Sufficient time must pass to allow the liquid level in the glass tubing to reach the liquid level in the drum. The top of the glass tube is capped with a stopper, or personnel with appropriate gloves may also cap it.

While capped, the glass tube is removed from the drum and the bottom end is inserted into the sample container. The stopper or hand is removed from the glass tube to allow its contents to fill, (one liter for analytical tests, 50 milliliters for fingerprint test), the sample container to approximately 90% of its capacity. If additional liquid is needed to fill the sample container, repeat the previous steps.

With gloves on, the glass tube shall be broken and placed into the drum. All used protective clothing shall be properly disposed of.

Spill containment equipment (e.g., sorbent, pads, vermiculite) must be readily available and easily accessible, if needed. If used to contain spillage during drum sampling activities, they shall be drummed and labeled with respect to their contents. The Shift Foreman shall notify management of any spill and subsequent measures to control it.

Sample collection activities require attention to detail and a routine that ensures quality and consistency while maintaining efficiency. The following should be performed or considered during each sampling event:

1. Before collection of samples, thoroughly evaluate the job.

2. Prepare all sampling equipment and sample containers prior to the job.

3. Place sample containers on flat, stable surfaces for receiving samples.

4. Collect representative samples and securely close containers as quickly as feasible.

5. Document all steps in the sampling process.

6. Minimize sampling-derived waste.

Sampling Methods

Drums and other closed containers will be opened using the manual method, which involves use of a non-sparking brass bung wrench to loosen the large or small bung plug on closed-head drums, or non-sparking wrenches to remove the nut(s) on open-head drums. Because drum bungs come in various shapes and sizes, an assortment of non-sparking sockets must be available to fit the bungs.

The sampling method will be determined by the type of container, access to (opening of) the container, and the physical state of the material in the container (liquid, sludge, or solid).

Liquid Waste: The sampler must carefully insert a length of glass tubing (drum thief or rod) through the drum opening to the drum bottom. For most liquids, a piece of tubing with an inside diameter of 6 to 8 mm is adequate, but a larger bore may be required for more viscous materials. The top end of the tube is raised from the drum to transfer the contents to the sample container. Removing the thumb or stopper allows the contents to empty into the container. The operation is repeated until adequate volume is obtained.

Sludge Waste: For sludges, a larger-bore glass tubing or a 40 mL VOA (volatile organics analysis) vial fastened to a length of wooden dowel may be used. The sampling apparatus may be discarded with other waste accumulated during the sampling operation.

Solid Waste: A disposable scoop may be used for an open-top drum, while a small ladle attached to a length of wooden dowel may be used to obtain material through a bung hole.

Sample Type: There are two general sampling techniques for defining waste and waste stream characteristics: grab and composite. A grab sample is defined as a discrete sample representative of a specific location at a given point in time. The sample is collected all at once and at one particular point in the sample medium.

Composite samples are non-discrete samples composed of one or more subsamples collected at various sampling locations and/or different points in time. Analysis of this type of sample produces an average value and can in certain instances be used as an alternative to analyzing a number of individual grab samples and calculating an average value.

FINGERPRINTING WASTE - GENERAL

Fingerprint analyses will be performed to ascertain that the waste meets the waste profile and shipping papers. Test instruments will be calibrated in accordance with manufacturer's recommendations.

The frequency for performing fingerprint tests is based on volume received at the HWMF as follows:

- 10% for waste streams of more than ten 55-gallon drums per year;
- 20% (or 2 drums) for waste streams of five to ten 55-gallon drums per year; and
- All containers except lab packs for waste streams of less than five 55-gallon drums per year.

An analysis pending label will be placed on drums identified for fingerprint tests.

Results of the tests will identify whether the container of waste is acceptable and passes or is non-conforming. If the drum passes, "pass" will be written on the drum and the data sheet and the pending label removed. If the drum is non-conforming, it will be moved to a work-in-process unit until discrepancies are resolved with the generator. If the drum is known to contain waste or a waste code not permitted for acceptance by the HWMF, the drum will be returned to the generator for appropriate management. Also, fingerprinting will then be performed on an additional 10% of the drums from that generator's waste stream. The additional testing will proceed until the Industrial Waste Specialist determines otherwise.

Fingerprint Test Procedures

An instrument used to detect Volatile Organic Compounds (VOCs), a Flame Ionization Detector (FID), or a Photoionization Device (PID), should be set up at the beginning of each drum sampling operation and is used to measure vapors evolved as drums are opened, prior to obtaining samples for testing. Portions of all phases present in the sample must be transferred to individual test tubes (each phase of each sample must be tested separately). Approximately 5 mL of liquid or 3 g of solid are required for testing.

An example of the Fingerprinting Data Sheet where fingerprint data are recorded is shown on page 105 and is completed according to the procedures outlined below. The first column of the form is Expected Profile Conditions for the parameters being tested and the remaining columns are for actual data observed during the fingerprint procedures.

1. Sample Description: While still in the sample jar, describe the physical nature of the sample. Include matrix (solid/liquid), number of phases and color.

2. Organic Vapors: Using a FID / PID open the drum and measure the concentration of organic vapors .

3. Corrosivity: Lay out the required number of pH test papers on a flat surface and place a small portion of each sample phase on a separate strip, using a spatula or eye dropper (if windy, dip paper). For solids, wet the paper with a few drops of water and apply the moistened paper to the solid. Read the pH indicated on the paper using the scale on the pH paper container for reference. If pH is 7 or more, check for cyanides and sulfides (see Steps 6 and 7). If pH is less than 7 or if water-soluble, check for the presence of oxidizers (see Step 5).

4. Water Reactivity: Add 3 mL of room temperature water to test tube, insert thermometer, note temperature, then add 1 ml of sample. Note the generation of heat in degrees F if vigorous reactions, bubbles, and/or vapors occur, indicating the sample is water reactive.

5. Oxidizers: The presence of oxidizing material contained in the sample is evaluated when the sample is water soluble.

a. Lay out the required number of potassium iodide starch papers (KI-starch papers) and acidify with one to two drops of 3Normal Hydrochloric Acid (3N HCL).

b. Apply a drop of liquid sample (or aqueous solid sample) to paper.

c. If paper turns blue or black after one to two minutes, sample is an oxidizer.

6. Sulfides: The sulfide content of a sample is generally performed only on samples with pH of 7 or more. The detection limit is 0.6 ppm of the sulfide ion.

a. Lay out the required number of lead acetate test papers and acidify with one to two drops of 3N HCL.

b. Apply a drop of liquid sample or touch to solid sample.

c. If paper darkens after one to two minutes, sample contains sulfides.

7. Cyanides: The presence of cyanides in a sample is generally evaluated only on samples with a pH of 7 or greater. Test kits are used that test for the presence of cyanide at 0-30 ppm levels. Instructions are provided in individual kits.

8. Chlorinated Hydrocarbons: The detection limit for this test is approximately 0.5% chlorine concentration as perchloroethylene. This test is performed on all samples that:

a. Are insoluble and have specific gravity greater than 1, or

b. Are slightly soluble and have Porta-FID reading greater than 10.

The procedure is as follows:

1. Heat copper wire in flame until flame is yellow, with no trace of green.
2. Cool wire by waving in ambient air for 10 -15 seconds.
3. Insert COOL wire in sample.
4. Insert sample-coated wire into flame.
5. A green flame indicates that chlorinated hydrocarbons are present.

HAZARDOUS WASTE MANAGEMENT FACILITY FINGERPRINTING DATA SHEET

HWMF Profile #:_________ Sample Date: ______ Profile Name:_________________________
Generator:_______________ Analyst:____________
Generator Cost Center:____ Notebook:__________

Expected Profile Conditions	Drum Number	Matrix Solid/ Liquid	Number of Phases	Color	Organic Vapor Detection	pH	Halogen Test	Water Reactive	Oxidizer	Sulfide	Cyanide	Comments
Test Results												
1												
2												
3												
4												
5												
6												
7												
8												
9												
10												

Waste Stream Re-evaluations

Waste streams should re-evaluated on an annual basis in order to maintain adequate knowledge of the waste. The annual re-evaluations will be based on operator knowledge and use of laboratory analytical results, as necessary. The re-evaluation may simply be a statement of no change to the waste stream by the generator.

CHAPTER 11

LAND DISPOSAL RESTRICTIONS NOTIFICATION

EPA has greatly restricted a generator's ability to dispose of hazardous wastes in landfills through regulations officially known as the Land Disposal Restrictions (LDR) and commonly known as the "land ban."

LDR paperwork must be completed by or for the generator and provided to the Treatment Storage and Disposal Facility (TSDF). The LDR paperwork should be submitted with or prior to every manifest (i.e., with the shipping facility manifest or during the profiling and characterization of the waste). Small quantity generators do not need to file this form with every manifest, but need to file once yearly for each waste stream.

INTRODUCTION

Hazardous and Solid Waste Amendments of 1984 (HSWA), Section 3004, includes restrictive provisions governing the land disposal of untreated hazardous wastes. HSWA required EPA to develop treatment standards stipulating concentrations or levels of hazardous constituents that are considered to be protective of human health and the environment, for all listed and characteristic hazardous wastes by May of 1990 (with a few exceptions, notable new TCLP D wastes).

Treatment standard concentrations are expressed either as:

1. concentrations of hazardous constituents in leachate produced from testing a sample of the waste using the TCLP (toxicity characteristic leaching procedure -- a toxicity test) ; or

2. concentrations of hazardous constituents present in the waste in weight percent (e.g., 2% lead). In addition to treatment standards expressed as concentrations, for a few wastes these standards are actually specified as a particular treatment technology (e.g., halogenated organic compounds -- incineration).

HSWA has divided the universe of listed and characteristic wastes into five groups and set schedules for EPA to develop treatment standards for these groups. The groups and schedules are:

<u>Solvents and Dioxins</u>: These were banned from land disposal (unless treated) effective November 8, 1986, and November 8, 1988, respectively.

<u>"California List" Wastes</u>: This group of hazardous wastes was originally developed by the State of California for their hazardous waste management program. It includes: <u>liquid</u> wastes containing certain metals, free cyanides, polychlorinated biphenyls, corrosives (pH less than 2.0) and certain wastes containing halogenated organic compounds. In addition, <u>solid</u> hazardous wastes containing halogenated organic compounds are also included in this group. The majority of these wastes was banned from land disposal (unless treated) effective July 8, 1987. Some wastes were given variances due to a lack of treatment capacity.

<u>"First-, Second-, and Third-third" Wastes</u>: The remaining list of listed and characteristic wastes was divided into thirds (see 40 CFR Part 268 for specific waste groupings). The "first third" wastes were banned effective August 8, 1988, and the "second third" June 8, 1989. The "third third" wastes were banned effective May 8, 1990.

<u>Newly Listed Wastes</u>: Additional wastes listed after November 8, 1984, will be evaluated on a case-by-case basis. EPA must make a determination of whether the waste may be land-disposed within six months of the identification or listing. However, HSWA does not impose an automatic prohibition on land disposal if EPA misses a deadline for a newly listed waste.

<u>Treatment, Storage, and Disposal Facility Requirements</u>

Specific land disposal restrictions requirements for TSDFs include:

- Ensuring compliance with generator recordkeeping requirements when residues generated from treating restricted wastes are manifested off-site; and
- Certifying that treatment standards have been achieved for particular wastes prior to disposal.

Facilities that generate land disposal restricted wastes need to ensure that the proper paperwork accompanies the manifest when restricted wastes are sent off-site for disposal. Facilities that store or treat restricted wastes or restricted waste residues, and send the residues off-site for disposal, are subject to the same recordkeeping regulations as generators. In addition, facilities that treat restricted wastes to the appropriate standard may send a certification with the manifest to the disposer verifying compliance. However, most disposal facilities generally test waste shipments to ascertain compliance with the treatment standards and to prepare their own certification.

Restricted wastes may be disposed in a hazardous waste landfill only if the waste meets the applicable treatment standard.

LAND DISPOSAL RESTRICTION REGULATIONS

Regulatory Overview

<u>Background</u>

The Hazardous and Solid Waste Amendments (HSWA) of 1984 established the Land Disposal Restrictions (LDR) Program. Congress pushed for LDR in response to a perception by the public that EPA's hazardous waste regulations were not doing enough to effectively protect human health and the environment. In Section 3004(m), Congress directed EPA to

> promulgate . . . levels or methods of treatment . . .which substantially diminish the toxicity of the waste or . . . the likelihood of migration of hazardous constituents . . . so that short-term and long-term threats to human health and the environment are minimized.

The regulations were gradually phased in from 1986 to May 1990.

<u>Scope of LDR</u>

LDR covers all RCRA hazardous waste. Newly identified hazardous wastes do not have LDR criteria. EPA is committed to develop LDR within 6 months after a waste becomes newly identified. In some cases the EPA fails to meet the schedule. Newly developed LDRs will be published in the Federal Register.

For the purpose of LDR, land disposal includes any placement of hazardous waste in a:

- Landfill
- Surface impoundment
- Waste pile

- Injection well
- Land treatment facility unit
- Salt dome or salt bed formation
- Underground mine or cave

Intent

The major intent of LDR is to ensure that:

- treatment standards are met prior to disposal;
- treatment is not evaded by long-term storage;
- there is actual treatment rather than dilution;
- recordkeeping and tracking follow a waste from cradle to grave;
- certification verifies treatment standard has been met.

Treatment standards apply not only to hazardous wastes, but also to:

- hazardous wastes mixed with solid wastes;
- wastes derived from hazardous waste;
- listed hazardous wastes mixed with non-solid wastes.

Use of Best Demonstrated Available Technology (BDAT)

All treatment must use BDAT, defined as follows.

Best: The technology must be based on performance data from well-designed and well-operated systems that substantially diminish the toxicity or mobility of hazardous constituents.

Demonstrated: Treatment technologies must be currently in use and preferably at a full-scale treatment facility.

Available: Technologies must be available for purchase or lease.

Other Technical Requirements

Compliance Monitoring

When a waste has a concentration-based standard, the treatment facilities measure compliance using either an extraction procedure (TCLP) or Total Waste Analysis (TWA), whichever is specified in the standard.

- TCLP is generally used for inorganic constituents when stabilization was used to develop the standard.

- TWA is generally used for organic constituents when a destruction or removal technology was used to develop the standard.

Testing frequency is specified in the Waste Analysis Plan (WAP).

There are no specific frequencies for analysis. In general, EPA likes to see each waste stream characterized annually, but they recognize that it is not practical for small quantity (a few drums per year) waste streams.

Storage Prohibition

Storage in a land-based unit is considered land disposal. Storage on-site in a tank or container is permissible to accumulate sufficient quantities necessary to facilitate proper recovery, treatment, or disposal. Wastes placed in storage prior to the effective dates of the LDRs are not subject to this prohibition.

Dilution Prohibition

In most cases, dilution is prohibited by RCRA. LDR dilution prohibition serves a twofold purpose: to ensure actual treatment of hazardous constituents, and to ensure appropriate treatment. Dilution to avoid an applicable treatment standard or effective date is impermissible. Prohibition does not preclude the addition of materials as part of a legitimate treatment process. Dilution of characteristic wastes is not permissible when:

- characteristic is due to toxicity;
- a specified method is required as the treatment standard (unless deactivation is the treatment standard).

Dilution of characteristic wastes is permissible when the waste is rendered non-hazardous in:

- a system that discharges to a Class 1 injection well;
- a system regulated under the Clean Water Act (unless a specified method is the treatment standard).

Aggregation is permissible dilution if:

- all prohibited wastes are amenable to the same treatment;
- treatment meets the appropriate standard.

Aggregation is not permissible when:

- "treated" by inappropriate methods or not treated at all;
- diluted to render the waste delistable.

National Capacity Variance

Treatment standards are immediately effective unless a national capacity variance is granted. National capacity variances are based on national treatment capacity. The new effective date for land disposal restrictions is based on the earliest date on which adequate capacity is available. Maximum extension is two years.

Use of LDR Notification to Determine the Treatment Method of Your Waste

Step 1 - Determine the proper EPA waste code.
Step 2 -Determine if waste is waste water (WW) or non-waste water (NWW). (It still could be a hazardous waste).

For F001-F005, the spent solvent wastes, WW is an aqueous mixture containing < 1% total solvents. Thus an aqueous mixture that is 98% water and 2% solvent is actually a NWW, not a WW.

For all other wastes, WW is an aqueous mixture containing < 1% TOC (total organic carbon or total suspended solids). Thus most solvent/water mixtures will be NWW.

Step 3 - Is there a specified technology standard?
Review 40 CFR Part 268 Table 2 to see if your waste code is listed. If it is, then use Table 1 to cross-reference the specified technology.

Step 4 - Is there a concentration-based treatment standard?
If the waste does not have a specified technology standard, go to the concentration-based tables. Using the waste code and the determination of a WW vs. NWW, find the concentration-based standard that applies to your waste.

INSTRUCTIONS FOR COMPLETING THE LDR NOTIFICATION

Although no standard form exists for the notification requirement, the following Sample Notification form is provided for guidance. If the profiled waste is found to have waste codes of F001, F002, F003, F004 or F005, place a "yes" in the first block on page one of the Land Disposal Restrictions Notification. On page two, "F" codes are listed separately with their associated constituents. Place a "yes" or "x" in the appropriate box to select the code, and identify the constituent(s) in the profiled waste stream by placing an asterisk or circling the constituent.

The next determination is whether the profiled waste contains any of the constituents or characteristics (pH) of a California List Waste. The California Waste Constituents and their regulated limits are on page three. If you have a constituent which exceeds the limit, place a yes in the space preceding the constituent and put the concentration in mg/L or mg/kg after the constituent. Remember to place a "yes" in the appropriate space on page one to indicate the inclusion of a California List Waste.

Example: If the profiled waste contains arsenic and has the associated waste code of D004, then it may also be a California List Waste if the arsenic concentration is greater than or equal to the regulatory limit of 500 mg/L.

If the profiled waste has waste code(s) other than F001-F005 or California List Wastes, place the EPA code(s) in the spaces under EPA Hazardous Waste Code. The only exception to this rule is if the waste contains new TCLP characteristic wastes which do not at this time have treatment standards.

Under the Treatability group, place wastewater or non-wastewater. Refer to the profile for the information. To complete the Subcategory, CFR reference for treatment standard, and five-letter code(s), check the following CFR references in order:

40 CFR § 268.41(a) -- Table CCWE
40 CFR § 268.42(a)(1)
40 CFR § 268.42(a)(2)
40 CFR § 268.42 -- Table 2
40 CFR § 268.42 -- Table 3
40 CFR § 268.42(c)
40 CFR § 268.43(a) -- Table CCW

Once the EPA waste code is located (cross referenced) with a valid subcategory which describes it and has the treatment standard for the waste (i.e., wastewater or non-wastewater), write the correct reference in the blank under CFR reference and fill in the last column. References are included in this section.

SAMPLE NOTIFICATION

Profile No.:_______________
Waste Stream Name:___________

LAND DISPOSAL RESTRICTION NOTIFICATION (per 40 CFR § 268.7)

The wastes identified on manifest number ________ and bearing the EPA Hazardous Waste Code(s) _______, _______, _______, _______, _______ are subject to the land disposal restrictions of 40 CFR Part 268. The wastes do not meet the treatment standards specified in Part 268 Subpart D or do not meet the prohibitions specified in § 268.32 or RCRA section 3004(d). The treatment standards or prohibition levels applicable to each waste are identified below:

(Put a "Y" in all blanks that apply)

____ This shipment includes F001-F005 spent solvents, as identified on the attached sheet. (If yes, complete page 2, place a "Y" in box(es) to identify individual constituents likely to be present in the waste.)

____ This shipment includes RCRA Section 3004(d) California List Wastes, as identified on page 3. (If yes, complete page 3 by placing a "Y" in box(es) to identify individual constituents likely to be present in the waste.)

____ This shipment includes additional wastes identified below:

	Hazardous Waste Code(s)	Treatability group (wastewater or non-wastewater)	Subcategory(1)	CFR reference treatment standard(2)	Five-letter code(s) (3)
1.	_______	___________	___________	___________	___________
2.	_______	___________	___________	___________	___________
3.	_______	___________	___________	___________	___________
4.	_______	___________	___________	___________	___________
5.	_______	___________	___________	___________	___________

(1) Refer to 40 CFR §§ 268.41 through 268.43(a) -- Table CCW.
Note: Not all waste code(s) have subcategory listings 268.42 -- Table 2 and Table 3; have subcategories. If the waste code(s) do not have subcategories, leave space blank or N/A.

(2) The CFR Reference must be one of the following:
40 CFR § 268.41(a); § 268.42(a)(1); (a)(2), (c) and Tables 2 and 3; or § 268.43(a) -- Table CCW.

(3) Five-letter code(s) are applicable for waste treatments found in: § 268.42 -- Tables 2 and 3.

Profile No.:_______________
Waste Stream Name:___________

TREATMENT STANDARDS FOR F001-F005 SPENT SOLVENTS

Instructions: Place a "Y" in the space beside each waste code included in the shipment; place a "Y" in spaces to identify the individual constituents likely to be present in each waste.

Hazardous waste description (Y,N)	Constituents of concern	Total composition, mg/kg	TCLP, mg/L	Wastewater, total composition, mg/L
___ F001-Spent halogenated solvents used in degreasing	_Carbon tetrachloride		0.96	0.05
	_Methylene chloride		0.96	0.20
	_Tetrachloroethylene		0.05	0.079
	_1,1,1-Trichloroethane		0.41	1.05
	_Trichloroethylene		0.091	0.062
	_1,1,2-Trichloro-1,2,2-triflouroethane		0.96	1.05
	_Trichlorofluoromethane		0.96	0.05
___ F002-Spent halogenated solvents	_Chlorobenzene		0.05	0.15
	_1,2-Dichlorobenzene		0.125	0.65
	_Methylene chloride		0.96	0.20
	_Methylene chloride (from pharmaceutical industry)			0.44
	_Tetrachoroethylene		0.05	0.079
	_1,1,1-Trichloroethane		0.41	1.05
	_1,1,2-Trichloroethane	*7.6*		0.030
	_Trichloroethylene		0.091	0.062
	_1,1,2-Trichloro-1,2,2-trifluoroethane		0.96	1.05
	_Tricholofluoromethane		0.96	0.05
___F003-Spent non-halogenated solvents	_Acetone		0.59	0.05
	_n-Butyl alcohol		5.0	5.0
	_Cyclohexanone		0.75	0.125
	_Ethyl acetate		0.75	0.05
	_Ethyl benzene		0.053	0.05
	_Ethyl ether		0.75	0.05
	_Methanol		0.75	0.25
	_Methyl isobutyl keytone		0.33	0.05
	_Xylene		0.15	0.05
___ F004-Spent non-halogenated solvents	_Cresols (& cresylic acid)		0.75	2.82
	_Nitrobenzene		0.125	0.66
___ F005-Spent non-halogenated solvents	_Benzene	*3.7*		0.070
	_Carbon disulfide		4.81	1.05
	_2-Ethoxyethanol	Incineration		Biological degradation or incineration
	_Isobutanol		5.0	5.0
	_Methyl ethyl ketone		0.75	0.05
	_2-Nitropropane	Incineration		(Wet oxidation or chemical oxidation) followed by carbon absorption; or incineration
	_Pyridine		0.33	1.12
	_Toluene		0.33	1.12

CALIFORNIA LIST CONSTITUENTS AND THEIR PROHIBITION LEVELS
(Identify constituents likely to be present)

	Waste Constituent	Regulatory Limit Concentration (Y,N)mg/L	Waste Concentration mg/L
	(Liquids Containing)		
______	Cyanides	1,000	____________
______	Arsenic	500	____________
______	Cadmium	100	____________
______	Chromium VI	500	____________
______	Lead	500	____________
______	Mercury	20	____________
______	Nickel	134	____________
______	Selenium	100	____________
______	Thallium	130	____________
______	Liquids with pH <= 2.0	---	____________
______	Liquids with PCBs	50 ppm	____________
______	Wastes (solids or liquids) containing HOCs*	1000 mg/kg	____________

A California List Waste is:

(A) A hazardous waste containing Halogenated Organic Compounds at a concentration of 1000 mg/kg or mg/L, or greater; or

(B) A liquid hazardous waste (including Free Liquids) having a pH equal to or less than 2.0 or containing any of the above listed metals.

* Halogenated Organic Compounds

CHAPTER 12

TRANSPORTATION

The EPA has adopted the Department of Transportation (DOT) regulations concerning the shipment of waste. The DOT's Hazardous Materials Regulations concerning naming, labeling, and marking of wastes, are explained in this chapter in addition to the use of the Uniform Hazardous Waste Manifest.

There are nine steps in preparing waste for shipment. The first step, hazardous waste determination, should have been completed during the profiling process. The remainder of the steps, in order, are: the selection of a proper DOT shipping name, hazard class, United Nations (UN) or North American (NA) number, packing group, markings and labels, placards, containers, and the completion of a manifest.

The proper shipping name, hazard class, UN/NA number, packing group, and necessary labels are presented in the "101" tables (49 CFR, § 172.101), Hazardous Materials Table. Reportable Quantities are the amount of material, which if spilled, must be reported to the appropriate authorities. The reportable quantity amounts can be found in the 101 appendix.

PREPARING FOR TRANSPORT

EPA has adopted by reference the DOT regulations regarding the transportation of hazardous waste found in 49 CFR Parts 172, 173, 178, and 179. These relate to packaging, labeling, marking, and placarding. In addition, EPA has specified a particular additional marking for hazardous waste containers (described below). A generator should consult with the transporter and the TSD facility prior to preparing hazardous waste for transport to ensure that the waste is properly packaged, labeled, marked, and placarded. A complete discussion of hazardous materials transportation regulations is beyond the scope of this book. The abbreviated discussion which follows applies generally to truck transport; transportation arrangements involving air or water transport are more complex and should be coordinated with the transporter.

General Discussion of DOT Regulations

To comply with the DOT regulations, the material being prepared for transport must be classified through the use of the DOT hazardous materials tables and instructions in 49 CFR § 172.101, and the proper shipping name and hazardous class identified. To use the table, the generator should know the major chemical constituents and characteristics of the waste, and should look up each of those in the table. The most specific category in the table should be used, with resort to general categories only where necessary (e.g., avoiding the n.o.s., not otherwise specified, category). For those mixtures with more than one constituent listed in the table, 49 CFR § 173.2 gives a simple rule for determining the appropriate shipping name and class. The 49 CFR § 172.101 table gives the proper shipping name, hazard class, DOT identification number, packing group, label(s), packaging requirement, and limitations on air or water shipment. For hazardous wastes, the proper shipping name should be prefixed with "Waste." The DOT shipping name, hazard class and number should be copied exactly. (Refer to § 172.101 table excerpts on page 121.)

EPA has prepared a guide and a series of inserts for small generators by industry class. This guide provides generator instructions and the insert gives the basic information obtainable from the DOT tables for many of the wastes which are produced by small generators in the industry class. These may be obtained from EPA through the hotline number (800-424-9346). The general publication number is EPA/530-SW-010; you must ask for the appropriate insert for your industry.

Selecting a Transporter

Unless you transport your waste yourself, you will need to carefully select a qualified transporter (see Schleifer, Jay, et al., *How to Comply With Hazardous Waste Laws, A Step-by-Step Guide,* Chapter Seven, "How to Select a Hazardous Waste Transporter"), especially since you are responsible for your waste even after it leaves your hands. The Department of Transportation audits all transporters and gives each a rating of Satisfactory, Conditional, and

Unsatisfactory. (A "Conditional" rating means that certain operations need to be improved, but the problems do not warrant a shutdown.)

Although you may not want to deal with transporters who receive "Conditional" or "Unsatisfactory" ratings, remember also that not all transporters rated "Satisfactory" will necessarily be right for you. In choosing a transporter, for example, you will want to consider your specific wastes and the transporter's scope of services and geographic location.

GENERAL INSTRUCTIONS FOR SELECTING PROPER SHIPPING NAMES

The Department of Transportation lists proper shipping names in 49 CFR § 172.101 -- The Hazardous Material Table. Although there are approximately 1700 proper shipping names, most hazardous waste generators will use only about 25. There is only one "best" shipping name which must be used.

Be cognizant of the following when selecting a proper shipping name:

1. You must use the most specific name available from the § 172.101 table. DOT considers the order of specificity to be:

   - Specific chemical name (e.g., "Hydrochloric Acid")
   - Chemical group or family (e.g., "Acid, liquid, n.o.s.")
   - End use of material (e.g., "Ink")
   - Generic End use (e.g., "Medicines, n.o.s.")
   - Hazard Class (e.g., "Corrosive liquid, n.o.s.")

2. The hazard ascribed to the name must actually be the hazard of the material (except for "+" materials). Otherwise a different name must be chosen.

3. You should not make additions or deletions to a shipping name except as explicitly allowed by the DOT, such as optional additional information. When you are shipping hazardous waste and the word waste is not included in the shipping name, you must add it.

4. When a n.o.s. (not otherwise specified) name is selected you must include additional information. Additional information should identify at least two major constituents present in the waste which contribute to the Hazard Class. If there is only one constituent in the waste, it should be identified.

5. Once you have found the best name, that name will be given a specific ID number [four digits preceded by UN or NA]. The three pieces of information you now have (the proper shipping name, the Hazard class, and the UN or NA number)are referred to as the US DOT "Basic Description."

6. The DOT description must be in the following order:
   - Proper shipping name, Hazard class, and UN or NA number
   - (additional information)

7. The packing group roman numeral ranks the degree of danger presented by the hazardous material. Packing group I indicates great danger; II, medium danger, and III, minor danger.

8. Reportable Quantity (RQ) values for the constituents in the material or waste can be found in the appendix to 49 CFR § 172.101, "List of Hazardous Substances and Reportable Quantities." RQ are also associated with EPA waste codes. If the weight of the constituent present in the container meets or exceeds the RQ, then RQ must be added to the proper shipping name. If an EPA waste code has an associated RQ value, consider the entire container weight as the basis for determining if an RQ must be included.

9 Final form for shipping hazardous waste when a packing group and a RQ quantity is present:

   - RQ, proper shipping name, hazard class, UN/NA number, packing group
   - (additional information)

Refer to 40 CFR § 172.101 appendix excerpts on page 122.

§172.101 Hazardous Materials Table

Symbols	Hazardous materials descriptions and proper shipping names	Hazard class or Division	Identification Numbers	Packing group	Label(s) required (if not excepted)	Special provisions	(8) Packaging authorizations (§173***)			(9) Quantity limitations		(10) Vessel stowage requirements	
							Exceptions	Non-bulk packaging	Bulk packaging	Passenger aircraft or railcar	Cargo aircraft only	Vessel stowage	Other stowage provisions
(1)	(2)	(3)	(4)	(5)	(6)	(7)	(8A)	(8B)	(8C)	(9A)	(9B)	(10A)	(10B)
	Dichlorodimethyl ether, symmetrical	6.1	UN2249	I	POISON	T25	None	201	243	Forbidden	Forbidden	D	40, 105
	1,1-Dichloroethane	3	UN2362	II	FLAMMABLE LIQUID	T7	150	202	242	5 L	60 L	B	40, M2
	1,2-Dichloroethane, see Ethylene dichloride												
	Dichloroethylene	3	UN1150	II	FLAMMABLE LIQUID	T14	150	202	242	5 L	60 L	B	
	Dichloroethyl sulfide	Forbidden											
	Dichloroisocyanuric acid, dry or Dichloroisocyanuric acid salts	5.1	UN2465	II	OXIDIZER	28	152	212	240	5 kg	25 kg	A	13
	Dichloroisopropyl ether	6.1	UN2490	II	POISON	T8	None	202	243	5 L	60 L	B	
	Dichloromethane	6.1	UN1593	III	KEEP AWAY FROM FOOD.	N36, T13	153	203	241	60 L	220 L	A	
	Dichloromonofluoromethane	2.2	UN1029		NONFLAMMABLE GAS.	B51	306	304	314, 315	75 kg	150 kg	A	
	1,1-Dichloro-1-nitroethane	6.1	UN2650	II	POISON	T8	None	202	243	5 L	60 L	A	12, 40, 48
	Dichloropentanes	3	UN1152	III	FLAMMABLE LIQUID	B1, T1	150	202	241	60 L	220 L	A	
	Dichlorophenyl isocyanates	6.1	UN2250	II	POISON		None	212	242	25 kg	100 kg	A	25, 40, 48
	Dichlorophenyltrichlorosilane	8	UN1766	II	CORROSIVE	A7, B2, B6, N34, T8, T26	None	202	242	Forbidden	30 L	C	40, M2
	Dichloropropane, see Propylene dichloride												
	1,3-Dichloropropanol-2	6.1	UN2750	II	POISON	T8	None	202	243	5 L	60 L	A	12, 40, 48
	Dichloropropene	3	UN2047	II	FLAMMABLE LIQUID	T8	150	202	242	5 L	60 L	A	M2
	Dichloropropene and propylene dichloride mixture, see Propylene dichloride												
	Dichlorosilane	2.3	UN2189		POISON GAS, FLAMMABLE GAS	2, B9, B13, B14, B31, B73	None	304	244	Forbidden	Forbidden	D	40
	Dichlorotetrafluoroethane	2.2	UN1958		NONFLAMMABLE GAS.		306	304	314, 315	75 kg	150 kg	A	

Table 1—Hazardous Substances Other Than Radionuclides

Hazardous Substance	Synonyms	Reportable Quantity (RQ) Pounds (Kilograms)
	1,1,1,2-Tetrachloroethane	
	1,1,2,-Tetrachloroethane	
Tetrachloroethene	Ethene, tetrachloro-	100 (45.4)
	Perchloroethylene *	
	Tetrachloroethylene *	
Tetrachloroethylene *	Ethene, tetrachloro-	100 (45.4)
	Perchloroethylene *	
	Tetrachloroethene	
2,3,4,6-Tetrachlorophenol	Phenol, 2,3,4,6-tetrachloro-	10 (4.54)
Tetraethyl lead *	Plumbane, tetraethyl-	10 (4.54)
Tetraethyl pyrophosphate *	Diphosphoric acid, tetraethyl ester	10 (4.54)
Tetraethyldithiopyrophosphate	Thiodiphosphoric acid, tetraethyl ester	100 (45.4)
Tetrahydrofuran *	Furan, tetrahydro-	1000 (454)
Tetranitromethane *	Methane, tetranitro-	10 (4.54)
Tetraphosphoric acid, hexaethyl ester	Hexaethyl tetraphosphate *	100 (45.4)
Thallic oxide	Thallium oxide T1203	100 (45.4)
Thallium ¢		1000 (454)
Thallium(I) acetate	Acetic acid, thallium(I+) salt	100 (45.4)
Thallium(I) carbonate	Carbonic acid, dithallium (I+)	100 (45.4)
Thallium(I) chloride	Thallium chloride T1C1	100 (45.4)
Thallium chloride TlCl	Thallium(I) chloride	100 (45.4)
Thallium(I) nitrate	Nitric acid, thallium(1+) salt	100 (45.4)
Thallium oxide T1203	Thallic oxide	100 (45.4)
Thallium selenite	Selenious acid, dithallium(1+) salt	1000 (454)
Thallium(I) sulfate *	Sulfuric acid, dithallium(I+) salt	100 (45.4)
Thioacetamide	Ethanethioamide	10 (4.54)
Thiodiphosphoric acid, tetraethyl ester	Tetraethyldithiopyrophosphate	100 (45.4)
Thiofanox	2-Butanone, 3,3-Dimethyl-1-(methylthio)-, O[(methylamino)carbonyl] oxime	100 (45.4)
Thioimidodicarbonic diamide [(H2N)C(S)]2NH	Dithiobiuret	100 (45.4)
Thiomethanol	Methanethiol	100 (45.4)
	Methyl mercaptan *	
Thioperoxydicarbonic diamide [(H2N)C(S)]2S2, tetramethyl-	Thiram	10 (4.54)
Thiophenol *	Benzenethiol	100 (45.4)
	Phenyl mercaptan @	
Thiosemicarbazide	Hydrazinecarbothioamide	100 (45.4)

Table 1—Hazardous Substances Other Than Radionuclides

Hazardous Substance	Synonyms	Reportable Quantity (RQ) Pounds (Kilograms)
Thiourea	Carbamide, thio- 10 (4.54)	
Thiourea, (2-chlorophenyl)-	1-(o-Chlorophenyl)thiourea	100 (45.4)
Thiourea, 1-naphthalenyl-	alpha-Naphthylthiourea	100 (45.4)
Thiourea, phenyl-	Phenylthiourea	100 (45.4)
Thiram	Thioperoxydicarbonic diamide [(H2N)C(S)2S2, tetramethyl-	10 (4.54)
Toluene *	Benzene, methyl-	1000 (454)
Toluenediamine *	Benzenediamine, ar-methyl-	10 (4.54)
Toluene diisocyanate *	Benzene, 1,3-diisocyanatomethyl	100 (45.4)
o-Toluidine	2-Amino-1-methyl benzene	100 (45.4)
p-Toluidine	Benzenaminew, 4-methyl-	100 (45.4)
o-Toluidine hydrochloride	Benzenamine, 2-methyl-, hydrochloride	100 (45.4)
Toxaphene *	Camphene	octachloro-
2,4,5-TP @	Propionic acid, 2-(2,4,5-trichlorophenoxy)-	100 (45.4)
	Silvex (2,4,5-TP)	
	2,4,5-TP acid	
2,4,5-TP acid	Propionic acid, 2-(2,4,5-trichlorophenoxy)-	100 (45.4)
	Silvex (2,4,5-TP)	
	2,4,5-TP @	
2,4,5-TP acid esters		100 (45.4)
1H-1,2,4-Triazol-3-amine	Amitrole	10 (4.54)
Trichlorfon		100 (45.4)
1,2,4-Trichlorobenzene		100 (45.4)
1,1,1-Trichloroethane *	Methyl chloroform *	1000 (454)
	Ethane, 1,1,1-trichloro-	
1,1,2-Trichloroethane	Ethane, 1,1,2-trichloro-	100 (45.4)
Trichloroethene	Trichloroethylene *	100 (45.4)
	Ethene, trichloro-	
Trichloroethylene *	Trichloroethene	100 (45.4)
	Ethene, trichloro-	
Trichloromethanesulfenyl chloride	Methanesulfenyl chloride, trichloro-	100 (45.4)
	Perchloromethyl mercaptan @	
Trichloromonofluoromethane	Methane, trichlorofluoro-	5000 (2270)
Trichlorophenol *		10 (4.54)
2,3,4-Trichlorophenol		
2,3,5-Trichlorophenol		
2,3,6-Trichlorophenol		

Rules for Determining the Proper Shipping Name for EPA Hazardous Waste

This system assumes the waste is contained in volumes of 55 gallons or less and shipped within the USA.

Rules:

1. Determine if the substance or mixture is hazardous. Is it:

   - Radioactive
   - Poison
   - Flammable
   - Combustible
   - Oxidizer
   - Corrosive
   - Irritant
   - Class 9
     - EPA Hazardous Waste by Characteristic
     - Listed EPA Waste
     - TSCA Waste

   Remember some DOT Hazards are defined by characteristic - e.g., Flash point 100 degrees F = Flammable liquid. Others are defined by listing - e.g., Blue Asbestos (crocidolite) and Brown Asbestos (amosite, mysorite) are Class 9.

2. If you do not have a specific chemical breakdown of the potentially hazardous waste, go through the suggested checklist to help select a primary hazard class. If you are unable to select a hazard class, the waste must be further characterized to find its components.

   Items to check the waste for:

   A. Flash Point:

      1. If it is $\leq$ 100 degrees F, then the waste falls into the DOT hazard class of a flammable and is an EPA ignitable waste.

      2. If it is > 100 degrees F, but < 140 degrees, then the waste falls into the EPA Hazardous Waste designation of an Ignitable and the DOT designation of combustible.

      3. If it is > 100 degrees, but < 200 degrees, then it falls into the DOT class of combustible.

   B. pH: If the pH of a solution is $\leq$ 2 or $\geq$ 12.5, it is an EPA and DOT corrosive.

   C. Does it contain a solvent? Is it halogenated? If yes to either question, it is a possible Flammable, Combustible, or Class 9.

   D. Does it contain an oxidizer, or is it reactive? If yes, it is a possible Oxidizer, Class 9.

E. Does it contain a RCRA metal? If yes, it is a possible Poison or Pesticide, Class 9.

F. Does it contain a TCLP organic? If yes, it is a possible Pesticide or Poison, Class 9.

3. If the substance or mixture possesses multiple hazard characteristics, use the following hierarchy to determine the primary class:

	Hazard Class
Radioactive	7
Poison A	2
Flammable Gas	2
Non-flammable Gas	2
Flammable Liquid	5
Flammable Solid	4
Corrosive Material (Liquid)	8
Poison B	6
Corrosive Material (Solid)	8
Irritating Material	6
Combustible Liquid, containers > 110 gallons	3
Combustible Liquid, containers < 110 gallons	3
Hazardous Waste n.o.s.	9

4. Note all possible names in 49 CFR 172.101 for your unknown, per the following priority:

   Technical names (See Dictionaries and Indexes); e.g., dimethyl ketone is "acetone"
   Chemical Generic (family) names; e.g., petroleum distillates are "Petroleum Oil, n.o.s."
   End Use of Material; e.g., "Paint"
   "n.o.s." End Use of material; e.g., "Insecticide, Liquid, n.o.s."
   DOT Class of Hazard; e.g., "Flammable Liquid, n.o.s."

   If found, note the proper shipping name, hazard class, labels required, packing group, packaging references, etc., as may be applicable. NOTE: Remember, if you find your material on DOT's List 172.101, that does not mean it is definitely regulated by the EPA under RCRA.

   CAUTION: Be certain your material in fact possesses the same hazard as listed in 172.101 under the DOT name you chose.

5. See if your material is listed as "Hazardous Substance" in the Appendix to 49 CFR 172.101, "List of Hazardous Substances and Reportable Quantities."

   Remember: Hazardous substance shipments require additional communications, including the letters "RQ" and, in some cases, the technical name of the hazardous substance (See Step #13).

6. See if your material is, or contains, a "listed waste" in 40 CFR Part 261, Subpart D. (See Sections 261.31 and 261.32 if it is a process waste, and Sections 261.33(e) and 261.33(f) if it is a chemical product.) If found, note the EPA Waste Code.

7. Does your material possess any of the characteristics defined in 40 CFR Part 261, Subpart C? If so, then it is a "Characteristics Hazardous Waste" (EPA). (See 40 CFR Sections 261.20 through 261.24 and State equivalents thereto.) Note EPA Waste Code.

8. If a proper shipping name was provided on Table § 172.101 per paragraph (4) above, then that also will be the proper shipping name for the waste. You must insert the word "waste" before the DOT proper shipping name if, and only if, the waste is required to be manifested by the US EPA. (Reference 49 CFR § 172.101(c) (10).)

9. If the material was not given a proper shipping name as per paragraph (4) above but it is considered a hazardous waste by the US EPA vis-a-vis steps in paragraphs (6), (7) and (8) above, then the proper shipping name is:

   "Hazardous waste, liquid or solid, n.o.s.", and the Hazard Class is "Class 9".

10. If the material is found to have a "Hazard Class n.o.s." proper shipping name, and the material has more than one kind of hazard, refer to the priority of hazard listing (step 3) name according to the highest priority hazard. Remember, in some cases, materials having more than one hazard may require multiple labeling, so refer also to 49 CFR § 172.402. The same is true for mixtures having more than one hazard. Additional descriptions on paper work may be required (49 CFR § 172.203).

11. If you have a specific chemical breakdown of the constituents for the waste, look up the chemical(s) or synonym(s) in CFR 49 Table § 172.101 to identify the proper shipping name(s). Synonyms will be found in Table § 172.101 Appendix in addition to Reportable Quantity (RQ) in pounds and kilograms. Record the shipping name(s) hazard class and RQ in pounds.

    Example:

    1. Name of Chemical (1) Haz Class ID # RQ
    2. Name of Chemical (2) Haz Class ID # RQ

12. If you only have a single component and its chemical name or synonym is listed in the Tables, preface the name with the word "waste."

    Example:

    1. Waste Name of Chemical Haz Class ID # RQ

13. If the RQ is $\leq$ 100 lbs., place RQ before the word "waste."

14. If the waste is composed of a single or multiple component(s) and you cannot find its chemical name or synonym in the 101 Tables, then look for the next most specific name from the example n.o.s. Table 1 list (see 49 CFR for complete list.)

N.O.S. Table 1

DOT ID #	Listed Shipping Names
NA1760	Acid, Liquid, n.o.s.
UN1987	Alcohol, n.o.s.
NA1719	Alkaline Liquid, n.o.s.
NA1133	Cement, n.o.s.
NA1790	Etching Acid, n.o.s.
NA2588	Insecticide, dry, n.o.s.
NA1993	Insecticide, liquid, n.o.s.
NA1270	Petroleum oil, n.o.s.
UN1078	Refrigerant Gas, n.o.s.
NA1325	Drug, n.o.s. (none)
NA1479	Drug, n.o.s. (oxidizer)
NA1993	Drug, n.o.s.. (flammable)
NA1479	Cosmetics, n.o.s.. (none)
NA1993	Cosmetics, n.o.s.. (flammable liquid)
NA1325	Cosmetics, n.o.s.. (flammable solid)
UN1461	Chlorate, n.o.s.
NA1461	Chlorate, wet, n.o.s.
NA1477	Nitrate, n.o.s.
UN1566	Arsenical Compound, Liquid, n.o.s.
UN1557	Arsenical Compound, Solid, n.o.s.
UN1566	Beryllium Compound, n.o.s.
NA1707	Thallium Salt, Solid, n.o.s.
NA1759	Cosmetics, Solid, n.o.s.
NA1759	Drugs, Solid, n.o.s.
UN1851	Medicines, Liquid, n.o.s.
UN1851	Medicines, n.o.s.
UN1851	Medicines, Solid, n.o.s.
UN1935	Cyanide Solution, n.o.s.
UN2025	Mercury Compound, Solid, n.o.s.
UN2757	Carbonate Pesticide, Liquid, n.o.s.
UN2757	Carbonate Pesticide, Solid, n.o.s.
UN2759	Arsenical Pesticide, Liquid, n.o.s.
UN2759	Arsenical Pesticide, Solid, n.o.s.
UN2761	Organochlorine Pesticide, Liquid, n.o.s.
UN2761	Organochlorine Pesticide, Solid, n.o.s.
UN2763	Triazine Pesticide, Liquid, n.o.s.
UN2763	Triazine Pesticide, Solid, n.o.s.
UN2765	Phenoxy Pesticide, Liquid, n.o.s.
UN2765	Phenoxy Pesticide, Solid, n.o.s.
UN2767	Phenylurea Pesticide, n.o.s.
UN2767	Phenylurea Pesticide, Solid, n.o.s.
UN2769	Benzoic Derivative Pesticide, Liquid, n.o.s.
UN2769	Benzoic Derivative Pesticide, Solid, n.o.s.
UN2771	Dithiocarbonate Pesticide, Liquid, n.o.s.
UN2771	Dithiocarbonate Pesticide, Solid, n.o.s.
UN2773	Phthalimide Derivative Pesticide, Solid, n.o.s.
UN2773	Phthalimide Derivative Pesticide, Liquid, n.o.s.
UN2775	Copper-based Pesticide, Solid, n.o.s.

UN2775	Copper-based Pesticide, Liquid, n.o.s.
UN2777	Mercury-based Pesticide, Liquid, n.o.s
UN2777	Mercury-based Pesticide, Solid, n.o.s.
UN2779	Substituted, Nitrophenal Pesticide, Liquid, n.o.s.
UN2770	Substituted, Nitrophenal Pesticide, Solid, n.o.s.
UN2781	Bipyridilium Pesticide, Liquid, n.o.s.
UN2781	Bipyridilium Pesticide, Solid, n.o.s.
UN2786	Organotin Pesticide, Liquid, n.o.s.
UN2786	Organotin Pesticide, Solid, n.o.s.
UN2813	Water-reactive Solid, n.o.s.
UN2845	Pyrophoric Liquid, n.o.s.
UN2918	Radioactive Material, Fissile, n.o.s.
NA9187	Organic Peroxide, Solid, n.o.s.
NA1993	Organic Peroxide, Liquid, n.o.s. (flammable)
NA1998	Organic Peroxide, Liquid, n.o.s. (corrosive)

15. If the waste is composed of a single or multiple components, and you cannot find a correct descriptive shipping name from the first Table of n.o.s. descriptions, then you might find a name from the second Table of n.o.s. (Please note that this list is not exhaustive. Refer to the DOT Tables for complete list.)

N.O.S. Table 2

DOT ID #	Listed Shipping Names
NA1993	Combustible Liquid, n.o.s.
NA1954	Compressed Gas, n.o.s. (Flammable)
NA1956	Compressed Gas, n.o.s. (Non-flammable)
UN1760	Corrosive Liquid, n.o.s.
UN2922	Corrosive Liquid, Poisonous, n.o.s.
UN1759	Corrosive Solid, n.o.s.
UN1078	Dispersant Gas, n.o.s. (Non-flammable)
NA1954	Dispersant Gas, n.o.s. (Flammable)
NA2814	Etiologic Agent, n.o.s.
UN1954	Flammable Gas, n.o.s.
UN2924	Flammable Liquid, Corrosive, n.o.s.
UN1993	Flammable Liquid, n.o.s.
UN1992	Flammable Liquid, Poisonous, n.o.s.
UN2925	Flammable Solid, Corrosive, n.o.s.
UN1325	Flammable Solid, n.o.s.
UN2926	Flammable Solid, Poisonous, n.o.s.
NA9188	Hazardous Substance, Liquid or Solid, n.o.s.
NA9189	Hazardous Waste, Liquid or Solid, n.o.s.
NA2814	Infectious Substance, Human, n.o.s.
NA1693	Irritating Agent, n.o.s.
NA1956	Non-flammable Gas, n.o.s.
NA1693	ORM-A, n.o.s.
NA1760	ORM-B, n.o.s.
NA9193	Oxidizer, Corrosive, Liquid, n.o.s.
NA9194	Oxidizer, Corrosive Solid, n.o.s.
NA9199	Oxidizer, n.o.s.
NA9200	Oxidizer, Poisonous, Liquid, n.o.s.

NA1953	Poisonous Liquid or Gas, Flammable, n.o.s.
NA1955	Poisonous Liquid, n.o.s.
UN2810	Poison B Liquid, n.o.s.
UN2928	Poisonous Solid, Corrosive, n.o.s.
UN2811	Poisonous Solid, n.o.s.
UN2811	Poison B, Solid, n.o.s.

16. Once the correct n.o.s. shipping name is identified, the correct form to use is:

    Form: Waste, Shipping Name n.o.s., Haz. Class ID # Packing Group
    Additional: (Name of primary constituent) (Name of second constituent)
    Labels: EPA Haz Waste, Hazard Class, if appropriate

NOTE:

- The shipping name of the primary constituent must be of the same hazard class as that of the shipping name. This does not depend on concentration of the constituent, as long as it is present in sufficient quantity to create the hazard classified.

- If the shipping name has only one hazard designation, but two or more constituents that contribute to the hazard class, list the two constituents with one as the primary, and one as the secondary -- in order of concentration.

- If one of the constituents is a poison in sufficient quantity to be poisonous, it must be listed as the primary or secondary constituent.

- If the shipping name specifies two hazards, the form to use is:

  Form: Waste, Haz (1) (physical state), Haz (2), n.o.s., Haz Class
  Additional: (Name of primary constituent) (Name of secondary constituent)
  Label: EPA Haz. Waste, Haz. Class, Haz. Class (2)

  To use this form, the waste must have at least one substance which generates two hazards or two substances, one of which creates one hazard and the second another hazard.

- If the substance presents an inhalation hazard, the package must be marked "Inhalation Hazard." This marking will also be included in labels for Haz. Class (2).

17. Note that it is required that a recognizable technical name be included in the proper shipping name for all poisons. If your material is a poison and has a non-technical proper shipping name such as an "end use" or "n.o.s." designation, then a technical name will also have to be inserted after the proper shipping name given by the table (and before the hazard class). Refer to 49 CFR § 172.203(m).

18. Empties: Any empty container that had previously contained a DOT hazardous material is to be considered still a hazardous material for shipping purposes. Unless it is appropriately "cleaned and purged" it must meet the requirements of 49 CFR §§ 173.24 and 173.28.

An empty is regulated by the EPA as a "Hazardous Waste" only as defined in 40 CFR § 261.33(c). A container that contained anything listed in Section 261.33(e) must be managed as described in § 261.33(c). The definition of when a container is "empty" under EPA rules can be found at 40 CFR § 261.7.

In any case, it is allowed, and it is recommended practice, to insert the phrase "RESIDUE: Last Contained" at the beginning of the proper shipping name used to ship and transport hazardous empties. (See 49 CFR § 172.203.)

19. For questions, call:
    DOT hotline...................... 202/366-4433

Example Shipping Names for F001 Wastes

Substance	RQ	Class/ Division	DOT #	Packing Group
Tetrachloroethylene (Syn: Perchloroethylene)	100	6.1	UN1897	II
Trichloroethylene	100	6.1	UN1710	III
Dichloromethane (Syn:Methylene Chloride)	1000	6.1	UN1593	III
1,1,1-Trichloroethane	1000	6.1	UN2831	III
Carbon Tetrachloride	10	6.1	UN1846	II

RULES:

1. If waste contains only one of the above-listed compounds, for the shipping name, preface the substance name with RQ (if the RQ is less than or equal to 100) and the word "waste." Labels are EPA Hazardous Waste.

   Format:

   Form: "RQ" waste, Name of Chemical, Class 6.1, UN#, Packing Group______

   Labels: EPA Haz. Waste.

   Example:

   Waste Dichloromethane, 6.1, UN 1593,III

2. If the waste contains the above substances in a mixture:

   Form: "RQ" Hazardous Waste, Liquid, n.o.s., Class 9, NA 9189, III
   Additional: (name of primary constituent) (name of secondary constituent)
   Labels: EPA Haz. Waste, Class 9

   Example:

   RQ Hazardous Waste, Liquid, n.o.s., Class 9, NA 9189, III

   (1,1,1-Trichloroethane) (Carbon Tetrachloride)

   Refer to Sample EPA Hazardous Waste Labels for additional examples

Example 1. DOT Labels and EPA Marking for Waste Flammable Liquid

HAZARDOUS WASTE

FEDERAL LAWS PROHIBIT IMPROPER DISPOSAL

IF FOUND, CONTACT THE NEAREST POLICE OR
PUBLIC SAFETY AUTHORITY OR THE
U.S ENVIRONMENTAL PROTECTION AGENCY

GENERATOR INFORMATION:

NAME ENVIRONMENTAL INFORMATION SERVICES, INC.

ADDRESS 4790 SHAWNEE PLACE, SUITE 102

CITY BOULDER STATE CO ZIP 80303

EPA ID NO. COD 983801598 EPA WASTE NO. D001, F003

ACCUMULATION (1) START DATE 6/29/93 MANIFEST DOCUMENT NO. 00001

Waste Flammable Liquid, n.o.s., 3, UN1993, I
(Dimethyl Formamide, Methanol)

D.O.T. PROPER SHIPPING NAME AND UN OR NA NO. WITH PREFIX

HANDLE WITH CARE!

Example 2. EPA Marking for Waste 1,1,1, Trichloroethane

HAZARDOUS WASTE

FEDERAL LAWS PROHIBIT IMPROPER DISPOSAL

IF FOUND, CONTACT THE NEAREST POLICE OR
PUBLIC SAFETY AUTHORITY OR THE
U.S ENVIRONMENTAL PROTECTION AGENCY

GENERATOR INFORMATION:

NAME ENVIRONMENTAL INFORMATION SERVICES, INC.

ADDRESS 4790 SHAWNEE PLACE, SUITE 102

CITY BOULDER STATE CO ZIP 80303

EPA ID NO. COD 983801598 EPA WASTE NO. F001, F002

ACCUMULATION (1) START DATE 6/29/93 MANIFEST DOCUMENT NO. 00001

Waste 1,1,1, Trichloroethane, 6.1, UN2831, III

(TCA)

D.O.T. PROPER SHIPPING NAME AND UN OR NA NO. WITH PREFIX

HANDLE WITH CARE!

Example 3. DOT Labels and EPA Marking for Waste Toluene Diisocyanate

HAZARDOUS WASTE

FEDERAL LAWS PROHIBIT IMPROPER DISPOSAL

IF FOUND, CONTACT THE NEAREST POLICE OR
PUBLIC SAFETY AUTHORITY OR THE
U.S ENVIRONMENTAL PROTECTION AGENCY

GENERATOR INFORMATION:

NAME ENVIRONMENTAL INFORMATION SERVICES, INC.

ADDRESS 4790 SHAWNEE PLACE, SUITE 102

CITY BOULDER STATE CO ZIP 80303

EPA ID NO. COD 983801598 EPA WASTE NO. U223

ACCUMULATION (1) START DATE 6/29/93 MANIFEST DOCUMENT NO. 00001

[RQ Waste Toluene Diisocyanate, 6.1, UN2078, II]

D.O.T. PROPER SHIPPING NAME AND UN OR NA NO. WITH PREFIX

HANDLE WITH CARE!

SHIPPING PAPERS

Hazardous waste shipments must be accompanied by shipping papers. Generally, the shipping paper takes the form of a manifest.

Hazardous Waste entries must include:

- The Proper Shipping Name;
- The Hazard Class, except where a hazard class shipping name is used;
- The UN or NA number;
- The total quantity of the material;
- "Emergency response information" must also be included with or on shipping papers.

PACKAGING

There are two approaches to packaging. First, the generator should check with the TSD facility to determine the packaging the TSD facility will accept or may prefer for the particular waste. The TSD facility normally will advise using a particular DOT specification packaging. The second approach is to ascertain the requirements from the DOT regulations. Most hazardous waste will be contained in open or closed head drums. Refer to example drum descriptions.

The specific section for packaging a particular material is found by reference to column 5 of the hazardous materials table at 49 CFR § 172.101. The DOT packagings of interest fall into the following categories:

- Limited Quantity: This category includes small packages. Typically, limited quantity packages are not above one to five gallons capacity, although some materials may not be allowed as limited quantity above a few ounces. Limited quantity packages are generally exempt from most of the specification packaging marking, labeling, and placarding requirements of DOT regulations. Laboratory samples would be shipped under this category.
- Fully regulated: Includes any other packaging allowed. Fully regulated packaging also may be divided into size classifications of "Bulk" (e.g., over 450 liters or 118.9 gallons, for liquids) and "Non-bulk."

DOT 17H

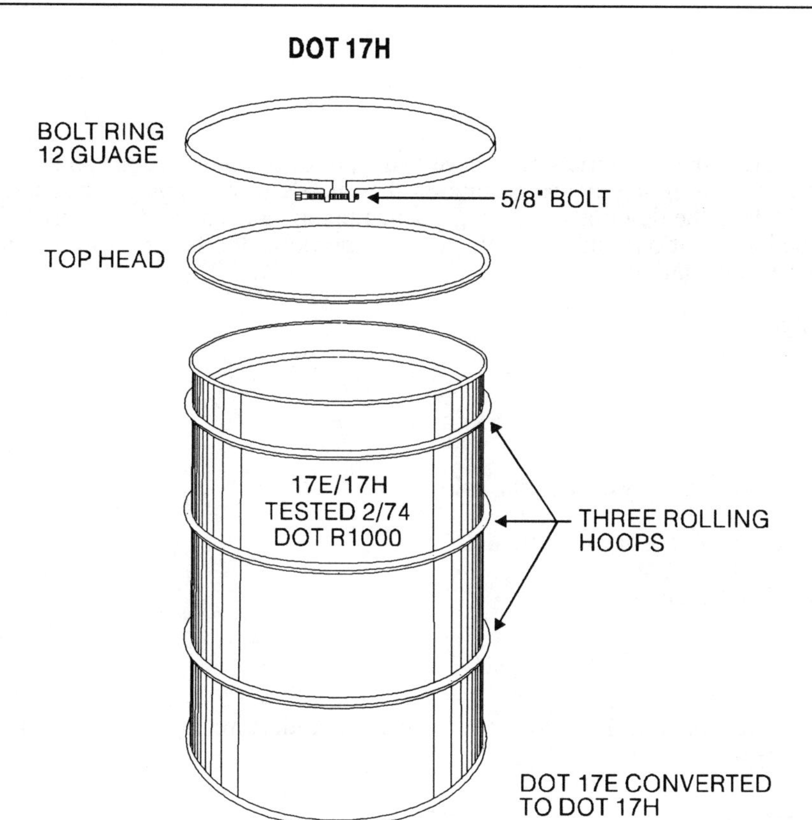

DOT 17E

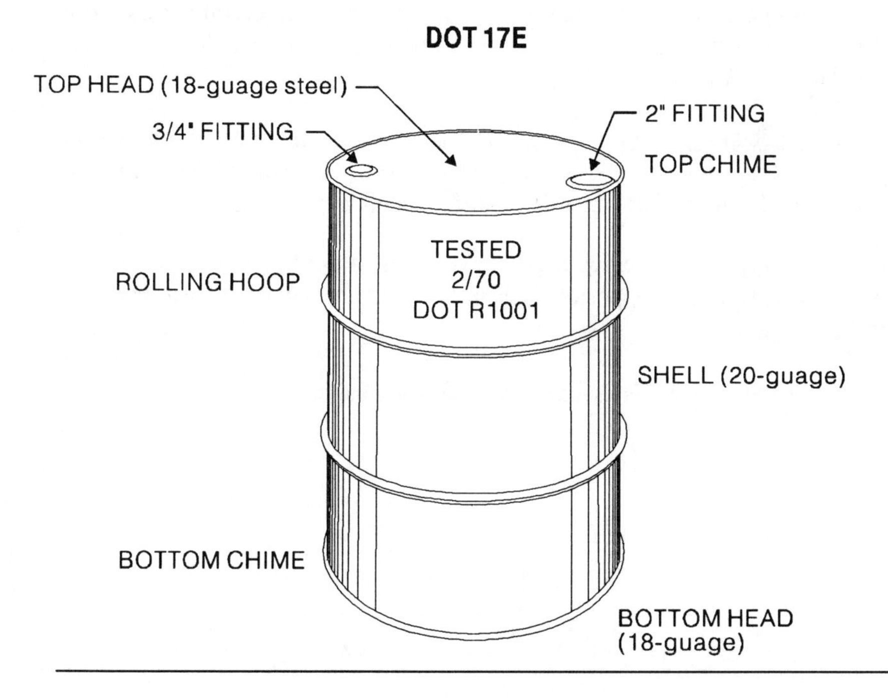

MARKING

Each package of hazardous materials must have two types of communications on the package, Markings and Labels. Marking means placing on the outside of a shipping container, one or more of the following: the descriptive name, proper shipping name, hazard class, identification number, instructions, cautions and/or weight. For our purposes the hazardous waste marking serves as the necessary marking.

Markings must be:

- durable;
- in English;
- on a background of sharply contrasting color;
- unobscured by labels or attachments; and
- away from other markings.

LABELING

Labels are placed on containers to provide an immediate indication of the hazards present or precautions needed.

Labels are of two types:

Hazard Class Labels, are diamond-shaped ("square on point") devices indicating the hazard class of the material being shipped (e.g., the "Flammable Liquid" label). In certain cases, both the primary hazard and a subsidiary hazard must be labeled (e.g., a flammable liquid which is also poisonous must be labeled for both hazard classes).

DOT also specifies certain Special Precaution Labels. These are labels intended to indicate an extra hazard or a special precaution to be taken during transportation. Examples include the "Dangerous When Wet" label for materials which have water-reactive properties.

The DOT Table (49 CFR § 172.101) specifies the type of label required. Labels should be attached to at least two sides, preferable opposite sides, of the container, with at least one near the opening point of the container.

Refer to hazardous material warning labels listed on the following page.

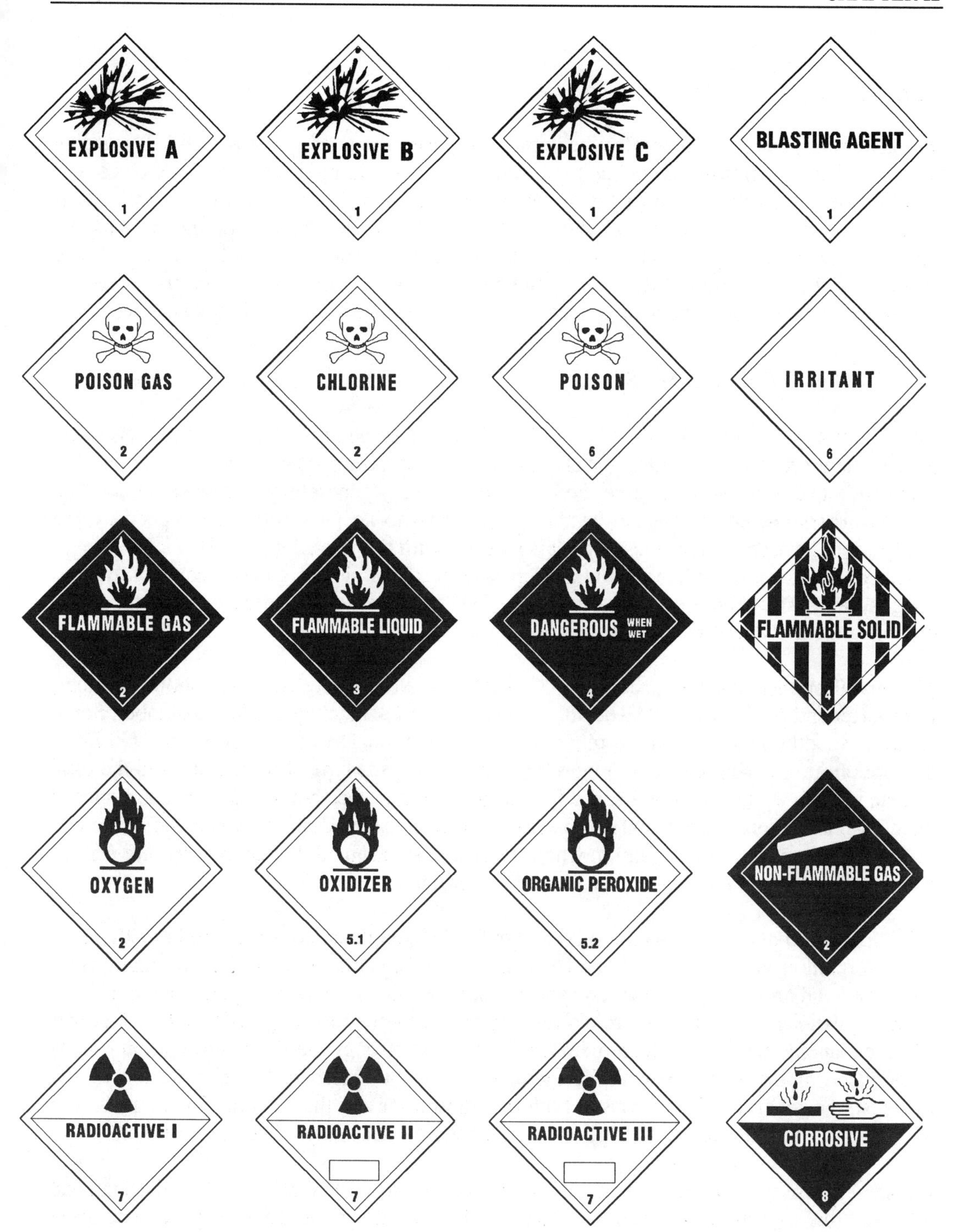

EXPLOSIVE A
1
EXPLOSIVE B
1
EXPLOSIVE C
1
BLASTING AGENT
1
POISON GAS
2
CHLORINE
2
POISON
6
IRRITANT
6
FLAMMABLE GAS
2
FLAMMABLE LIQUID
3
DANGEROUS WHEN WET
4
FLAMMABLE SOLID
4
OXYGEN
2
OXIDIZER
5.1
ORGANIC PEROXIDE
5.2
NON-FLAMMABLE GAS
2
RADIOACTIVE I
7
RADIOACTIVE II
7
RADIOACTIVE III
7
CORROSIVE
8

PLACARDING

The generator must placard or offer the initial transporter the appropriate placards. Placarding is required for all transportation of hazardous waste with limited exceptions. A placard is required on each side or end of the trailer or vehicle in which the hazardous materials is transported; the front placard may be attached to the tractor instead of the trailer. The placards must meet size, durability, color, and other requirements and must be securely affixed to the vehicle. Most transporters and TSD facilities can advise the generator on correct placarding given the proper shipping name(s). Most transporters provide their own placards.

PREPARING THE MANIFEST

The manifest system (40 CFR §§ 262.20-262.23) is the keystone of the RCRA hazardous waste management program. Any time you ship any amount of hazardous waste you are required to complete a manifest (with one exception for small quantity generators, discussed below). The manifest is completed and signed by the generator, who keeps a copy; it is signed by the transporter and accompanies the waste; it is received with the waste by the TSD facility, where it is checked against the shipment and discrepancies resolved, signed, and one copy returned to the generator. The generator, transporter, and TSD facility are all required to maintain copies of the manifest as a record of the shipment.

The manifest includes the generator US EPA ID No., name, address, phone number, transporter and transporter EPA ID No., the designated TSD facility and its ID number. Some states have additional information requirements. In section 11 of the manifest the US DOT description (proper shipping name) must be entered, then the type of containers, and the total volume of each waste. There is a space for the generator to specify for each waste any special handling instructions. The manifest form also includes a certification which the generator makes regarding waste minimization when the manifest is signed. States may require certain additional information if they require use of their own manifests.

It is important that the information on the manifest be completed accurately and exactly, since this information is required by both EPA and DOT regulations as well as by many state environmental and transportation laws and regulations. All of the above agencies also enforce these regulations and there also are discrepancy checks by the receiving TSD facility. An error which commonly results in enforcement action is failure to copy the DOT information exactly as provided in the DOT manuals. If there is any doubt regarding the preparation of the manifest, the generator should consult both the transporter and the TSD facility for assistance and follow up with the appropriate government agency as necessary. (See example manifest.)

For small quantity generators, there is one exemption (40 CFR § 262.20(e)) to the manifest rules and this regards waste which is reclaimed. This alternative procedure requires a contract between the small quantity generator and the recycling facility. The contract must specify the type of waste and the frequency of shipments, and a copy must be retained by the generator for at least three years following termination of the agreement. The reclaimer of the waste also must be the transporter of the waste and any regenerated material returned to the generator.

UNIFORM HAZARDOUS WASTE MANIFEST

1. Generator's US EPA ID No.	Manifest Document No.	2. Page 1	Information in the shaded is not required by Federal
C O D 9 3 8 1 5 9 8	0 0 0 0 1	1 of 1	

3. Generator's Name and Mailing Address
Environmental Information Services, Inc.
4790 Shawnee Place, Suite 102
Boulder, CO 80303

4. Generator's Phone (303) 494-4067

A. State Manifest Document Number

B. State Generator's ID

5. Transporter 1 Company Name: ABC Transport
6. US EPA ID Number: C O D 9 3 0 0 0 0 0 0 0

C. State Transporter's ID 40195

D. Transporter's Phone 303-494-40

7. Transporter 2 Company Name
8. US EPA ID Number

E. State Transporter's ID

F. Transporter's Phone

9. Designated Facility Name and Site Address
Rollins OPC
5756 Alba Street
Los Angeles, CA 90058

10. US EPA ID Number: C A D 0 5 0 8 0 6 8 5 0

G. State Facility's ID

H. Facility's Phone

	HM	11. US DOT Description *(Including Proper Shipping Name, Hazard Class and ID Number)*	12. Containers No.	Type	13. Total Quantity	14. Unit Wt/Vol	I. Waste
a.	X	Waste Dichloromethane, 6.1, UN 1593, III	0 0 1	D M	0 0 0 5 0	G	F001 F002
b.	X	Waste Flammable Liquid, N.O.S., 3, 1993, I	0 0 3	D M	0 0 1 5 0	G	D001 F003
c.	X	Waste 1,1,1 Trichloroethane, 6.1, UN 2831, III	0 0 1	D M	0 0 0 5 0	G	F001 F002
d.	X	RQ Waste Toluene Diisocyanate, 6.1, UN 2078,II	0 0 1	D M	0 0 0 5 0	G	U223

J. Additional Descriptions for Materials Listed Above

K. Handling Codes for Wastes Listed Abo

15. Special Handling Instructions and Additional Information
Note: Toluene Diisocyanate is a poison, Handle With Care.
For emergency or spill call: CHEMTREC, 1-800-424-9300
Refer to Emergency Response Guidebook, 11a.74, 11b.27, 11c.74, 11d.57

16. **GENERATOR'S CERTIFICATION:** I hereby declare that the contents of this consignment are fully and accurately described above by proper shipping name and are classified, packed, marked, and labeled, and are in all respects in proper condition for transport by highway according to applicable international and national government regulations.

If I am a large quantity generator, I certify that I have a program in place to reduce the volume and toxicity of waste generated to the degree I have determined economically practicable and that I have selected the practicable method of treatment, storage, or disposal currently available to me which minimizes the prese future threat to human health and the environment; **OR**, if I am a small quantity generator, I have made a good faith effort to minimize my waste generation and the best waste management method that is available to me and that I can afford.

Printed/Typed Name	Signature	Month Day
James L. Lieberman		

17. Transporter 1 Acknowledgement of Receipt of Materials

Printed/Typed Name	Signature	Month Day

18. Transporter 2 Acknowledgement of Receipt of Materials

Printed/Typed Name	Signature	Month Day

19. Discrepancy Indication Space

20. Facility Owner or Operator: Certification of receipt of hazardous materials covered by this manifest except as noted in Item 19.

Printed/Typed Name	Signature	Month Day

GENERATOR

TRANSPORTER

FACILITY

NIFORM HAZARDOUS WASTE MANIFEST
(Continuation Sheet)

21. Generator's US EPA ID No.	Manifest Document No.	22. Page	Information in the shaded areas is not required by Federal law.

Generator's Name

L. State Manifest Document Number

M. State Generator's ID

Transporter ___ Company Name	25. US EPA ID Number	N. State Transporter's ID
		O. Transporter's Phone
Transporter ___ Company Name	27. US EPA ID Number	P. State Transporter's ID
		Q. Transporter's Phone

HM	US DOT Description *(Including Proper Shipping Name, Hazard Class, and ID Number)*	29. Containers No.	29. Containers Type	30. Total Quantity	31. Unit Wt/Vol	R. Waste No.

dditional Descriptions for Materials Listed Above

T. Handling Codes for Wastes Listed Above

Special Handling Instructions and Additional Information

Transporter ___ Acknowledgement of Receipt of Materials		Date
Printed/Typed Name	Signature	*Month Day Year*
Transporter ___ Acknowledgement of Receipt of Materials		**Date**
Printed/Typed Name	Signature	*Month Day Year*

Discrepancy Indication Space

Instructions for Uniform Hazardous Waste Manifest (EPA Forms 8700-22 and 8700-22A)

Federal regulations require generators and transporters of hazardous waste and owners/operators of hazardous waste treatment, storage, and disposal facilities (TSDFs) to use Form 8700-22 (and, under certain circumstances, Continuation Sheet 9700-22A) for both *inter- and intrastate* transportation. Use of the form in relation to *international* shipments is noted below under item 18. (Federal regulations do not require the information requested in shaded areas, but states may require generators and TSDF owners/operators to complete some of those items as part of their own state manifest reporting requirements.)

Read all instructions before completing the form(s). The forms have been designed for use on a 12-pitch (elite) typewriter. Pen may be used-- select one with a firm point and press hard.

Enter the required information in the proper spaces on the form(s) in accordance with the following instructions.

Form 8700-22:

Items 1-16 are to be filled out by generator, items 17 and 18 by transporter, item 20 by the facility owner or operator or designated agent.

1. Generator's US EPA ID No./Manifest Document No. -- Enter the 12-digit US EPA identification number, and the 5-digit identification number uniquely assigned to this specific manifest by the generator.
2. Page 1 of ___ -- Enter the total number of pages used to complete the manifest; that is, this page plus the number of Continuation Sheets (8700-22A), if any.
3. Generator's Name and Mailing Address -- Use the address of the location that will handle the returned manifest forms.
4. Generator's Phone Number -- Enter the number of a phone at which an authorized agent may be reached in the event of an emergency.
5. Transporter 1 Company Name -- Enter the company name of the transporter that will transport the waste first.
6. US EPA ID Number -- Enter the 12-digit identification number assigned to the transporter identified in item 5.
7. Transporter 2 Company Name -- If a second transporter will transport the waste, enter that company name here.
8. US EPA ID Number -- If a second transporter is listed in item 7, enter that company's 12-digit identification number here.

   Note: Every transporter used between the generator and the designated facility (see item 9) must be listed, in the order in which they will transport the waste. If there are more than two, their company names and 12-digit identification numbers must be listed on one or more Continuation Sheets, each of which has spaces (items 24-27) for two additional transporters.

9. Designated Facility Name and Site Address -- Enter the company name and site address (which may be different from the mailing address) of the facility designated to receive the waste listed on this manifest.
10. US EPA ID Number - Enter the 12-digit identification number of the facility identified in item 9.
11. US DOT Description, etc. -- Enter the proper shipping name, hazard class, and ID number (UN/NA) for each waste as identified in 49 CFR Parts 171-177.

    Note: This form has room for four separate wastes; if there are additional wastes to be described, use Continuation Sheet (s).
12. Containers (No. and Type) -- For each waste listed in item 11, enter the number of containers and the appropriate abbreviation from the accompanying Table for type of container.
13. Total Quantity -- For each waste listed in item 11, enter the total number of units of measure.
14. Unit Wt./Vol. -- For each waste listed in item 11, enter the unit of measurement using the appropriate abbreviation from the accompanying table.
15. Special Handling Instructions and Additional Information -- Generators *may* use this space to indicate any special transportation, treatment, storage, disposal, or bill-of-lading information. For international shipments, destined for treatment, storage, or disposal outside U.S. jurisdiction, generators *must* enter the point of departure (city and state) here. States *may not* require use of this space for any additional, new, or different information.
16. Generator's Certification -- The generator must read, sign (by hand), and date the certification statement. If a mode *other than* highway is used, the word "highway" should be lined out and the mode used (rail, water, air) should be inserted in the space at the end of the item box. If a mode *in addition* to the highway mode is used, enter that additional information (e.g. "and rail") in the space at the end of the box.

    Generators may reprint the words "On behalf of" in the signature block, or may hand-write this phrase in the signature block prior to signing the certification.

    Primary exporters shipping hazardous wastes to a facility located outside of the United States must add to the end of the first sentence of the certification the following words:

    "and conforms to the terms of the EPA Acknowledgment of Consent to the shipment."

    In signing the certification statement, those generators who have not been exempted by statute or regulation from the duty to make a waste minimization certification under section 3002 (b) of RCRA are also certifying that they have complied with the waste minimization requirements.

17. Transporter 1 Acknowledgment of Receipt -- Enter the name of the person accepting the waste on behalf of the first transporter. That person must acknowledge acceptance of the waste described in the manifest by signing and entering the date of receipt.

18. Transporter 2 Acknowledgment of Receipt -- name, date, and signature of person accepting the waste on behalf of the second transporter, if any. (If there are more than two transporters, acknowledgments will be entered on Continuation Form(s), at items 33 and 34.)

    Note: For *international* shipments, transporter responsibilities are as follows:

    Exports: Transporter must sign and enter the date the waste left the United States, in item 15.

    Imports: Transporters of RCRA-regulated hazardous wastes into the United States from another country are responsible for completing the manifest, which must accompany the shipment at the point of entry.

19. Discrepancy Indication Space -- If there is any significant discrepancy between any waste described on the manifest and the waste actually received at the designated (or alternate) facility, that discrepancy must be noted here by the authorized representative of the facility owner/operator.

    Owners and operators of facilities located in states not authorized to administer the hazardous waste management program who cannot resolve significant discrepancies within 15 days of receipt of the waste must send a copy of the manifest at issue, along with a letter describing the discrepancy and attempts to reconcile it, to the EPA Regional Administrator.

    Owners and operators of facilities located in states that have been authorized by the EPA to administer the program should contact the appropriate state agency for information on the state's Discrepancy Report requirements.

    Note: For lists of EPA Regional offices and states authorized to administer the hazardous waste management program, see Chapter 4 of this book.

Form 8700-22A:

This form must be used if more than two transporters will be involved in transporting the waste, or if more than four different wastes are included in the shipment. Information in shaded areas is not required by federal regulation, but some or all of it may be required as part of state manifest reporting requirements.

21. Enter generator numbers as they appeared in item 1 on the first page of the manifest.

22. Enter page number of the Continuation Sheet (the first one used will be page 2).

23. Same as item 3 from page 1.

24. Company name of additional transporter, preceded by the "order" number. Because there is room for two additional transporters on each continuation sheet, item 24 will be used for Transporter 3 (or 5, or 7, etc.).

25. Enter the 12-digit identification number of the transporters named above.

26. Transporter 4 (or 6, or 8, etc.) and company name.

27. Above-named transporter's 12-digit ID number.

28. For items 28-32, see instructions for items 11-15.

33. Enter same number used in item 24 to indicate order of transporter (3,5,7, etc.). Then follow instructions for item 17.

34. Enter same number used in item 26 to indicate order of transporter (4, 6, 8, etc.). Then follow instructions for item 18.

35. See instructions for item 19.

Container-Type and Unit-of-Measure Abbreviations for Uniform Hazardous Waste Management

Type of Container	
BA = Bags, made of burlap, cloth, paper, or plastic.	DF = Drums, barrels, kegs made of fiberboard or plastic.
CF = Cartons, cases, boxes (including roll-offs) made of fiber or plastic.	DM = Drums, barrels, kegs made of metal.
CM = Cartons, cases, boxes made of metal.	DW = Drums, barrels, kegs made of wood.
CW = Cartons, cases, boxes made of wood.	TC = Tank cars.
CY = Cylinders.	TP = Tanks, portable
	TT = Cargo tanks (tank trucks)

Unit of Measure	
G = Gallons (liquids only)	L = Liters (liquids only)
P = Pounds	K = Kilograms
T = Tons (2,000 pounds)	M = Metric Tons (1,000 kilograms)
Y = Cubic yards	N = Cubic meters

SUMMARY: STEPS TO TRANSPORT HAZARDOUS MATERIALS

1. Classification: Is it a Hazardous Material or a Hazardous Waste? What are the hazard(s)?
2. Naming: Selecting the most specific proper shipping name available.
3. Picking a Package: Usually 55-gallon drum -- 17E closed-head, 17C or 17H open-head drum.
4. Marking and Labeling: Marking requirements, Hazard Class labels, Hazardous Waste Marking.
5. Preparing Shipping Papers: Complete manifest.
6. Placarding: Placard transport vehicle if appropriate. Placard for flammable, corrosive, etc.
7. Loading, Moving and Unloading: Load and brace containers correctly.
8. Incidents & Emergencies: Emergency Preparedness ("Emergency Response Communications"). DOT written reports, DOT emergency reports, EPA reports.

HAZARDOUS MATERIALS INCIDENT REPORTS

The Hazardous Materials Incident Report, form DOT 5800.1, is used to report hazardous materials incidents by all modes of transportation. A detailed written report is required whenever there is an unintentional release of a hazardous material during transportation (or temporary storage related to transportation). This report is also required to be submitted for any quantity of hazardous waste and reportable quantities of hazardous substances discharged during transportation.

NOTE: Single copies are available from the Research and Special Programs Administration by sending a mailing label to: Training Branch DHM-51, Training and Technical Assistance Division, Research and Special Programs Administration, US DOT, Washington DC 20590.

Multiple copies of the Hazardous Materials Incident Reports may be purchased from the organizations listed below:

American Trucking Association, Inc.
Customer Service Section
2200 Mill Road
Alexandria, VA 22314
703/838-1754

UNZ and Company
190 Baldwin Avenue
Jersey City, NJ 07306
800/631-3098; 202/795-5400;
212/432-1205 (NewYork)

J. J. Keller
145 West Wisconsin Avenue
Neenah, WI 54956

DEPARTMENT OF TRANSPORTATION Form Approved OMB No. 2137-0039

HAZARDOUS MATERIALS INCIDENT REPORT

Instructions: Submit this report in duplicate to the Manager, Information Systems, Materials Transportation Bureau, Department of Transportation, Washington, D.C. 10590 (ATTN: DMT-63). If space provided for any item is inadequate, complete that item under Section H, "Remarks", keying to the entry number being completed. Additional copies in this prescribed format may be reproduced and used, if on the same size and kind of paper.

A INCIDENT

1. TYPE OF OPERATION

1☐ AIR 2☐ HIGHWAY 3☐ RAIL 4☐ WATER 5☐ FREIGHT FORWARDER 6☐ OTHER (Identify) ______

2. DATE AND TIME OF INCIDENT (Month - Day - Year) ______ a.m. ______ p.m.

3. LOCATION OF INCIDENT

B REPORTING CARRIER, COMPANY OR INDIVIDUAL

4. FULL NAME

5. ADDRESS (Number, Street, City, State and Zip Code

6. TYPE OF VEHICLE OR FACILITY

C SHIPMENT INFORMATION

7. NAME AND ADDRESS OF SHIPPER (Origin address)

8. NAME AND ADDRESS OF CONSIGNEE (Destination address)

9. SHIPPING PAPER IDENTIFICATION NO.

10. SHIPPING PAPERS ISSUED BY ☐ CARRIER ☐ SHIPPER ☐ OTHER (Identify) ______

D DEATHS, INJURIES, LOSS AND DAMAGE

DUE TO HAZARDOUS MATERIALS INVOLVED

11. NUMBER OF PERSONS INJURED

12. NUMBER OPERSONS KILLED

13. ESTIMATED AMOUNT OF LOSS AND OR PROPERTY DAMAGE INCLUDING COST OF DECONTAMINATION (Round all in dollars) $

14. ESTIMATED TOTAL QUANTITY OF HAZARDOUS MATERIALS RELEASED

E HAZARDOUS MATERIALS INVOLVED

15. HAZARD CLASS (Sec. 172.101, Col.3)	16. SHIPPING NAME (Sec. 172.101, Col.2)	17. TRADE NAME

F NATURE OF PACKAGING FAILURE

18. (Check all applicable boxes)

☐	(1) DROPPED IN HANDLING	☐	(2) EXTERNAL PUNCTURE	☐	(3) DAMAGE BY OTHER FREIGHT
☐	(4) WATER DAMAGE	☐	(5) DAMAGE FROM OTHER LIQUID	☐	(6) FREEZING
☐	(7) EXTERNAL HEAT	☐	(8) INTERNAL PRESSURE	☐	(9) CORROSION OR RUST
☐	(10) DEFECTIVE FITTINGS, VALVES, OR CLOSURES	☐	(11) LOOSE FITTINGS, VALVES OR CLOSURES	☐	(12) FAILURE OF INNER RECEPTACLES
☐	(13) BOTTOM FAILURE	☐	(14) BODY OR SIDE FALIURE	☐	(15) WELD FAILURE
☐	(16) CHIME FAILURE	☐	(17) OTHER CONDITIONS (Identify)		19. SPACE FOR DOT USE ONLY

Form DOT F 5800.1 (10-70) (9/1/76)

G	PACKAGING INFORMATION - If more than one size or type packaging is involved in loss of material show packaging information separately for each. If more space is needed, use Section H "Remarks" below keying to the Item number.						
	ITEM				#1	#2	#3
	20	TYPE OF PACKAGING INCLUDING INNER RECEPTACLES (Steel drums, wooden box, cylinder, etc.)					
	21	CAPACITY OR WEIGHT PER UNIT (55 gallons, 65 lbs., etc.)					
	22	NUMBER OF PACKAGES FROM WHICH MATERIAL ESCAPED					
	23	NUMBER OF PACKAGES OF SAME TYPE IN SHIPMENT					
	24	DOT SPECIFICATION NUMBER(S) ON PACKAGES (21P, 17E, 3AA, etc., or none)					
	25	SHOW ALL OTHER DOT PACKAGING MARKINGS (Part 178)					
	26	NAME, SYMBOL, OR REGISTRATION NUMBER OF PACKAGING MANUFACTURER					
	27	SHOW SERIAL NUMBER OF CYLINDERS, CARGO TANKS, TANK CARS, PORTABLE TANKS					
	28	TYPE DOT LABEL(S) APPLIED					
	29	IF RECONDITIONED	A	REGISTRATION NO. OR SYMBOL			
		OR REQUALIFIED, SHOW	B	DATE OF LAST TEST OF INSPECTION			
	30	IF SHIPMENT IS UNDER DOT OR USCG SPECIAL PERMIT, ENTER PERMIT NO.					

H REMARKS - Describe essential facts of incident: Including but not limited to defects, damage, probable cause, stowage, action taken at the time discovered, and action taken to prevent future incidents. Include any recommendations to improve packaging, handling, or transportation of hazardous materials. Photographs and diagrams should be submitted when necessary for clarification.

31. Name of person preparing report (Type or print)	32. Signature
33. Telephone No. (Include Area Code)	34. Date Report Prepared

Reverse of Form DOT F 5800.1 (10-70)

CHAPTER 13

ESTABLISHING A RECORDKEEPING SYSTEM

Recordkeeping is an integral responsibility of the RCRA administrator. It is one of the fundamental hazardous waste management requirements. The administrator must be able to demonstrate compliance with agency requirements. In addition it serves as a source of information for waste minimization and compliance/disposal costs. All records must be kept for at least three years, but you would be well advised to keep them in perpetuity.

It is advised to keep a file of all manifest and exception reports in a safe and secure location. All analytical results/ data should be filed with the waste profile they support.

This chapter provides guidance for recordkeeping and reporting.

INTRODUCTION

In Chapter 9 we discussed the information needed to complete the hazardous waste stream profile. A correctly completed profile will contain all of the information needed for generators to collect, label, store and ship their waste. In addition it will provide necessary information for the offsite treatment, storage and disposal company. If the profiles are verified by a waste analysis plan evaluation at least once a year, they will be kept current and should satisfy regulatory officials.

Every completed profile package should contain:

1. A copy of the profile signed and dated by the generator and approved and dated by the RCRA administrator.

2. An acceptance or rejection letter addressed to the generator clearly stating whether the generator may or may not transfer the waste to the onsite permitted storage facility. An acceptance letter tells the generator that the facility is permitted to accept his waste for storage. A rejection letter states that the onsite permitted storage facility does not have the necessary or appropriate permit to store the particular waste. In that case, the waste must be held by the generator until a coordinated pickup can be arranged to take the waste to an offsite TSD facility.

3. Example DOT and EPA labels are provided to simplify the completion of the labeling by the generators.

4. A waste manifest if the waste is hazardous and must travel over the public highway to reach the onsite permitted storage facility, or be shipped to an offsite TSD facility.

5. A Bill of Lading for non-hazardous waste or waste that can be transferred to the oversight storage facility without using the public highway.

6. A copy of the correct DOT Emergency Response Guide page that can be provided to the shipper of the waste.

7. For hazardous waste a completed Land Disposal Restriction Notification form.

8. A profile description that contains an overview of the most important information contained in the profile. Some states require such a form.

9. May contain a non-mandatory cover letter written by the person who completed the profile listing logic for completing the profile and recommendations. This is a good practice when employing consultants.

A completed profile package should be provided to RCRA administrator, to the generator, and to the responsible supervisor at the onsite permitted storage facility. If there is an environment compliance officer(s) who is(are) responsible for a group of generators, he should also be provided with a copy if the profiled waste stream is from one of his generators.

The profile packages should be filed by numerical order. A cross reference file should be easily accessible that contains:

1) the profile number; 2) whether the onsite permitted storage facility may accept the waste for storage; 3) whether waste is hazardous or non-hazardous; 4) EPA compatibility group; 5) waste stream name; 6) generator; 7) technical content and 8) approval date. This reference should be kept updated as additional waste profiles are added.

It soon becomes obvious that a computer database is not only convenient for the storage of the profile information, but that updates can be made quickly and easily and printed out for distribution. If a database is created, it is important that it be password protected so that unauthorized users can not change or delete information.

The author has created an integrated database for waste profiling; it assists the user in the completion of the profile and allows for the printing of any or all of the documents contained in the waste profile package. The program was created using dBase III+. It can be run on IBM compatible 286, 386 or 486 computers. It is strongly recommended that large generators (100 or more profiles) use a 386 or 486 machine.

The particular program created assists the profiler in selecting EPA Codes, DOT Shipping Names, DOT Guide Numbers, CASRN Numbers, RQs and Land Disposal Restricted Form information such as subcategory, CFR reference and treatment codes. The program works this way: the user inputs basic information.; if there is a question about specific pieces of information requested, the user may go to a help screen which presents a list of selections to choose from; the user selects the items by highlighting them and pressing the return key to have requested information posted in the block.

RECORDKEEPING AND REPORTING

The recordkeeping and reporting requirements for generators provide EPA and the states with a method to track the quantities of waste generated and the movement of hazardous wastes. The generator regulations in 40 CFR Part 262 contain three primary recordkeeping and reporting requirements:

- Biennial reporting;
- Exception reporting; and

- Three year retention of reports, manifests, and test records.

Biennial Reporting

Large quantity generators who transport hazardous waste off-site must submit a biennial report to the Regional Administrator by March 1 of each even-numbered year. The report details the generator's activities during the previous calendar year and includes:

- EPA identification number and name of each transporter used throughout the year;
- EPA identification number, name, and address of each off-site treatment, storage, or disposal facility to which waste was sent during the year;
- Quantities and nature of the hazardous waste generated;
- Efforts made to reduce the volume and toxicity of the wastes generated; and
- Changes in volume or toxicity actually achieved, compared with those achieved in previous years..

Generators who treat, store, or dispose of their hazardous waste on-site also must submit a biennial report that contains a description of the type and quantity of hazardous waste the facility handled during the year, and the method(s) of treatment, storage, or disposal used.

Exception Reports

In addition to the biennial report, generators who transport waste off-site must submit an exception report to the Regional Administrator if they do not receive a copy of the manifest signed and dated by the owner or operator of the designated facility within 45 days from the date on which the initial transporter accepted the waste. The exception report must describe efforts made to locate the waste, and the results of those efforts. Small quantity generators who do not receive a copy of the signed manifest from the designated facility within 60 days must explain the exception on a copy of the original manifest and send it to the Regional Administrator.

Three Year Retention of Reports, Manifests, and Test Records

The generator must keep a copy of each biennial report and any exception reports for a period of at least three years from the date the report was submitted. The generator also is required to keep a copy of all manifests for three years or until he or she receives a copy of all manifests signed and dated from the owner or operator of the designated facility. The manifest from the facility must then be kept for at least three years from the date on which the hazardous waste was accepted by the initial transporter. Finally, the records of the waste analyses and determinations undertaken by the generator must be kept for at least three years from the date that the waste was last sent to an on-site or off-site TSDF. The periods of retention mentioned above can be extended automatically during the course of any unresolved enforcement action regarding the regulated activity or as requested by the Administrator.

LAND DISPOSAL RESTRICTIONS

One of the major impacts HSWA has had on the implementation of the RCRA program is the restriction on land disposal for certain hazardous wastes. HSWA Section 3004 restricts the land disposal of hazardous waste beyond specified dates unless the wastes are treated according to treatment standards. These requirements are found in 40 CFR Part 268. The land disposal restrictions are referred to as the "land ban."

Generator Requirements

The land disposal restrictions affect both generators and TSDFs (see Chapter 11 for details on the land disposal restrictions requirements). Generators are responsible for:

- Determining whether their waste is restricted from land disposal; and
- Ensuring that all manifested shipments of restricted wastes are accompanied by the proper records when sent to a TSDF.

Recordkeeping Requirements

All shipments of restricted hazardous wastes manifested off site must be accompanied by the proper records. The type of records depends on whether or not there is a treatment standard specified for a particular waste. Recordkeeping requirements for restricted wastes with treatment standards differ from those for restricted wastes without standards. These requirements are addressed separately below.

Restricted Wastes with Treatment Standards. If a generator determines that the restricted waste exceeds the applicable treatment standard, a notification must accompany the manifest to the treatment facility and must include:

- EPA hazardous waste number;
- Applicable treatment standards;
- Manifest number of the waste shipment; and
- Waste analysis date, if available.

If the generator determines that his or her restricted waste meets the treatment standard without the need for further treatment, the generator must supply the same notification as well as a certification stating that the waste complies with the treatment standard. The notification and certifications must accompany that waste shipment to the hazardous waste disposal facility.

Restricted Wastes Without Treatment Standards. If the generator determines that treatment standards are not developed for the restricted waste, the waste may be land-disposed only if the generator prepares demonstration(s) and certification to accompany the manifest. The demonstration(s) and certification attest that land disposal is the only viable alternative to manage the waste. Chapter 11 outlines the paperwork requirements for all land disposal restricted wastes.

Additional Generator Requirements and Exclusions. Additional generator requirements apply to persons who export their wastes or submit their waste for land disposal. In addition, farmers have been excluded from complying with generator requirements under certain circumstances.

INTERNATIONAL SHIPMENTS

Prior to HSWA, hazardous wastes could be exported from the United States with only minimal notice to EPA or the receiving country. HWSA set additional notification requirements for such exports. These requirements prohibit the export of hazardous waste unless the exporter obtains prior written consent from the receiving country. This statement must be attached to the manifest accompanying each waste shipment.

To export a hazardous waste, the Administrator must first be notified by the exporter 60 days prior to when the waste is scheduled to leave the United States. This notification must be completed only for the first shipment in any 12-month period, unless basic information about the nature and frequency of the shipments changes. If the importing country agrees to accept the hazardous waste, EPA will send an "Acknowledgment of Consent" to the exporter, who may then export the waste to the importing country. Recordkeeping and reporting requirements are similar to those for domestic shipments of hazardous waste.

CHAPTER 14

ESTABLISHMENT OF TRAINING PROGRAMS

Training programs are an integral part of a cost effective waste management program. Some training is required by OSHA, and other training by the EPA and DOT. The RCRA administrator should ensure that all of his generators take a facility specific generator training course. This course should be short and to the point. It should give the generators the necessary information to accumulate, package and label their waste properly. Employees who work at a permitted storage treatment or disposal facility must take at least a 24 hour OSHA course with annual 8 hour refresher courses. Workers handling hazardous substances must have a hazard communication training course. Additional training is advised for site specific operations.

INTRODUCTION

This chapter will endeavor to outline and describe the necessary and recommended training courses and topics associated with hazardous waste management activities; they include:

- 24 hour OSHA training course per 29 CFR § 1910.120;
- 16 hours of on-the-job site specific OSHA/RCRA training;
- 8 hour OSHA refresher training annually;
- Respirator use and care training;
- Emergency response training; and
- Hands-on fire extinguisher training (optional).

There are a relatively large number of companies providing 24 hour OSHA training.

RCRA/OSHA COMPLIANCE PROGRAM

Introduction

An RCRA / OSHA Compliance Program should include the requirements of 29 CFR § 1910.120 for facilities that store hazardous waste. Paragraph (p)(1) requires a written safety and health program covering employees involved in hazardous waste operations. The program must identify, evaluate and control safety and health hazards at a Hazardous Waste Management Facility (HWMF) to provide for employee protection. The safety and health program must address emergency response procedures for the HWMF. If HWMF employees will evacuate the work site location in the event of an emergency and will not assist in handling the emergency, then an Emergency Action Plan satisfies the requirement. An Emergency Action Plan must be included in the Safety and Health Plan. If a Hazardous Material Response Team is not contracted to cover emergency response, then significant employee training must be conducted.

Training

All employees involved in hazardous waste operations at a controlled HWMF must complete an Initial Training which provides for a minimum of 24 hours of classroom training covering the following topics:

- Medical Surveillance
- Chemical Hazard Recognition
- Physical Hazard Recognition
- Personal Protective Equipment
- Respiratory Protection
- Drum Handling
- Decontamination Procedures
- Contingency Plan
- Review of Safety and Health Program
- Initial Spill Response
- Hazard Communication

In addition to the 24-hour classroom training, the employee must have a minimum of 8 hours of on-the-job-training. Written documentation of the training must be maintained. (Usual cost of training is between $375 and $450 per employee.)

Subsequent to the initial training, HWMF employees must receive 8 hours of refresher training annually. The annual training should cover the topics presented in the initial training. Special training must be provided to employees operating special or heavy equipment. All training must be given by qualified instructors experienced in the subject matter being discussed.

Safety and Health Plan

The Safety and Health Plan must cover the following areas:

- Chemical and Physical Hazard identification;
- Control Measures including site monitoring, personal protective equipment, and respiratory equipment;
- Emergency Action Plan including emergency procedures, escape procedures, duties, and employee and manager training.

Hazardous Communication Program

The purpose of a Hazard Communication Program is to inform employees of the existence of possible hazards in the workplace when handling toxic or otherwise dangerous materials. The primary topics covered are:

- Labels and other forms of warning;
- Material Safety Data Sheets (MSDSs);
- Employee information and training.

Medical Surveillance Program

The Medical Surveillance Program must provide for:

- Medical examination scheduling;
- Medical examination content;
- A physician's written opinion concerning employee capability;
- A recordkeeping system; and
- Employee responsibilities.

Decontamination Program

The Decontamination Program must cover:

- Personnel decontamination;
- Equipment decontamination;
- Inspection of equipment;
- Disposal of contaminated supplies and equipment.

Material Handling Program

The Material Handling Program must cover:

- Waste handling procedures (i.e., drum handling);
- Safe use of equipment.

New Hazards Program

The New Hazards Program must set up a system for training employees about new hazards introduced into the work environment.

HAZARDOUS WASTE GENERATOR TRAINING

It is highly advantageous to provide hazardous waste generator training. It provides the RCRA administrator with an opportunity to meet with generators, outline the facility's plan for handling hazardous waste, clarify generators' responsibilities, and answer generators' questions.

Course Outline for Hazardous Waste Generator Training

The following Hazardous Waste Generator Training Outline is designed for a course that can be delivered in about two hours.

1. Introduction
   - Who are the players?
   - What are the relevant regulations?
   - What must generators do with their hazardous waste?
   - What should generators learn from this course?

2. Responsibilities
   - Company responsibilities
   - Employee/generator responsibilities
   - Identify and label all waste
   - Profile all waste suspected of being hazardous
   - Properly manage waste at generator site
   - Keep drum inventory log

3. Proper Waste Accumulation
   - Satellite accumulation areas
   - 90-day accumulation areas

4. Satellite Accumulation Areas
   - No time limit
   - 55-gallon or 1-quart of acutely hazardous waste limit
   - Must be located at the point of generation
   - Must be under the control of the operator/generator employee

5. 90-Day Accumulation Areas
   - 90-day storage limit
   - No volume limit

6. Spill Prevention and Preparedness
   - Contingency Plan and Emergency Procedures

7. Personnel Training Requirements
   - Hazard Communication training

8. Closure Requirements
   - For 90-day areas

9. Container Management
   - Selected DOT shipping containers should be in good condition
   - Select containers that will be compatible with the waste
   - Keep containers closed except when adding or removing waste
   - Handle containers so as to avoid rupture or leakage
   - Inspect accumulation areas weekly
   - Store ignitable or reactive waste more than 50 feet from the property line
   - Separate incompatible wastes
   - Do not mix wastes; put each waste into the correct drum

10. Proper DOT Shipping Containers
    - DOT open-head drums (17C or 17H)
    - DOT closed-head drums (17E or 17F)

11. Drum Labeling and Storage Requirements
    - Label container as hazardous waste
    - Mark container with date as soon as volume limit is reached (gallons for hazardous waste, and 1 quart for acutely hazardous waste)
    - Immediately move full containers to 90-day or permitted storage areas
    - Provide adequate aisle space at storage areas
    - Move waste held at a 90-day storage facility to a permitted facility or TSDF prior to the expiration of the 90-day holding period
    - Inspect waste held in 90-day storage areas weekly for leaking or damaged drums

12. Examples of Hazardous Waste Labels
    - Review each entry on label

13. Review Generator Responsibilities
    - Safely generate, store and ship waste
    - The need to follow procedures carefully
    - Put waste into correct drum -- don't mix wastes
    - Notify supervisor of problems

14. Review Step-by-Step Waste Management Practices, and Satellite and 90-day Accumulation rules
    - Profile waste
    - Use proper DOT drum
    - Label drum
    - Add waste and then close container
    - Fill out drum inventory log each time waste is added
    - Manage spills and leaks
    - Keep track of volume in containers as well as in satellite accumulation areas
    - Print the date on the label of a container when it is full and move it to a 90-day or Permitted
    - Arrange for transport of full drums
    - Prepare drums for transport

- Conduct weekly inspections
- Post sign-in area
- Maintain records

15. Preparing Drums for Transport
    - Fill out shipping papers
        - manifest, or
        - Bill of lading
    - Attach copy of inventory log
    - Sign shipping papers
    - Arrange for transport
    - Do not store drums on loading dock

16. Generator Recordkeeping
    - Department-specific procedures
    - Container inventory logs
    - Bill of lading
    - Inspection logs
    - Corrective actions
    - Employee training records
    - Documentation of the implementation of emergency procedures

CHAPTER 15

SOLID WASTE MANAGEMENT UNIT INVESTIGATION PLAN

If a facility decides on obtaining a part B permit one of the requirements will be a solid waste management unit (SWMU) investigation plan to uncover or discover locations and operations that have historically generated waste or hazardous waste. It is something like a witch hunt, because what the generator believes is a solid waste management unit and what regulators determine was or is a unit may be quite different. The goal of the investigation is to identify and then remediate past contamination. The investigation is similar to a Phase I & II environmental assessment of real property conducted according to the ASTM standard E 1527 -93.5

INTRODUCTION

The purpose of a Solid Waste Management Unit (SWMU) Investigation Plan is to provide a description of the information gathering and technical investigation activities which will be implemented to identify and investigate releases of hazardous wastes or hazardous waste constituents from SWMUs at a site. For purposes of this plan, the site includes contiguous properties to the Hazardous Waste Management Facility (HWMF).

The plan is divided into three sections which correspond to the following major activities:

- Identification of SWMUs, including historic and present uses;
- Characterization of SWMUs using existing information, including identification of physical and chemical characteristics; and
- Field investigations, as necessary to determine if there has been a release of hazardous wastes or hazardous waste constituents from a SWMU.

The plan will serve as the basis for development of SWMU-specific work plans. Each SWMU-specific work plan will identify the individual(s) responsible for the implementation of the work plan and for compliance with any agency requirements associated with the SWMU.

SOLID WASTE MANAGEMENT UNIT IDENTIFICATION

The first task in implementing an SWMU investigation is to gather and evaluate existing information regarding active and inactive SWMUs at the site. The following sections describe the planned activities to gather and evaluate this information.

Aerial Photographs

Aerial photographs will be obtained and reviewed. The photographs will be reviewed for indications of any solid waste management facilities, and to help determine the period of operation of any SWMUs which are identified.

Interviews

Interviews with site personnel (and former employees, to the extent possible) who may be knowledgeable about past and present waste management activities will be conducted. These interviews will focus on identification of waste handling practices, potential areas where wastes may have been managed, active waste management areas, and waste characteristics. The aerial photographs and site maps will be used to help personnel locate known or potential SWMUs.

Field Survey

A field survey will be conducted to follow up on information which was collected during the aerial photograph review and interviews. The field surveys will primarily focus on visual location of SWMUs (where possible), and identification of SWMU site conditions, including:

- Potential physical or chemical safety hazards;
- Topography and ground cover;
- Visual indications of contamination (staining, stressed vegetation, etc.);
- Environmental considerations, such as proximity to drainages and supply wells; and
- Access restrictions.

Detailed field notes and photographs will be taken during field survey to document site conditions.

SOLID WASTE MANAGEMENT UNIT CHARACTERIZATION

Once a SWMU or a potential SWMU has been identified, all available information concerning the wastes and the SWMU location will be collected and evaluated. The following activities will be conducted to determine, to the extent possible, the physical and chemical characteristics of each SWMU.

Additional Interviews

Additional interviews will be conducted with site personnel who currently work or have worked in the vicinity of each potential SWMU. These interviews are intended to supplement previous interviews and the field surveys. Major objectives include identification of the originating process (including materials used) and the characteristics of any wastes, identification of time periods of waste management activities, and a description of any other relevant information which may exist.

Record Review

All pertinent facility and public records will be reviewed for each potential SWMU. The record review is intended to identify the types and characteristics of wastes which may have been managed at each SWMU, to quantify where possible the amount of each waste type managed at each SWMU, to determine whether hazardous wastes were or are managed at each SWMU, and to obtain any available construction information for each SWMU. The types of records which will be reviewed include:

- Raw material descriptions and usage (where applicable);
- Process flow diagrams;
- Utility drawings related to waste streams;
- Waste analyses;

- Waste disposal information;
- SARA Title III reports;
- Construction drawings;
- Logbooks and correspondence;
- Previous site investigations (if any);
- Publications concerning soils, geology and hydrogeology of the area;
- Permits or other agency submittals; and
- Spill or emergency report information.

Characterization Summary

All of the existing information will be evaluated and summarized to provide a physical and chemical characterization of each potential SWMU. This summary will be used to identify a SWMU and determine whether there are any immediate hazards posed by any SWMU, to determine whether hazardous wastes are or were ever handled at the SWMU, determine significant data gaps, describe any known releases and plan field investigation activities where needed. The SWMUs will be prioritized based on potential hazards and the status of any ongoing activities at or related to the SWMU. Field investigations will be initiated based on the priority ranking.

FIELD INVESTIGATION

Field investigations will be conducted in areas where there is a reason to believe that there has been a release of hazardous waste or hazardous waste constituents from a SWMU. These field investigations will be conducted at SWMUs where hazardous wastes were or may have been handled, and will be scheduled based on priority ranking. The field activities may vary at each SWMU as a result of site or waste characteristics and whether there have been any known releases.

The field investigation activities will be conducted in phases. The first phase (Phase I) will consist of field investigations to determine if there has been a release of hazardous waste or hazardous waste constituents from the SWMU. As such, the Phase I investigations will be limited to characterization of the waste material and investigation of the first potentially affected media. In most cases the first potentially affected media will be soil. If it is determined that there has not been any release, no further field investigations will be conducted. However, if there are indications of a release, either as a result of evaluating existing data or the Phase I investigation, the necessity of additional phases of investigation will be evaluated to determine the nature and extent of the release. The scope of these additional investigations is not addressed in this plan.

The following sections describe the types of field investigation activities which may be conducted; however, not every activity will necessarily be conducted at each SWMU. SWMU specific work plans will be developed which will provide details of field investigation techniques, analytical methods, sample locations, quality assurance, data evaluation methods and reporting.

Waste Characterization

If existing information is insufficient to locate either a known or suspected SWMU, an attempt will be made to determine the location of the SWMU using surface geophysical methods, soil vapor surveys, drilling or excavation. One or more of these methods may be used, depending on the waste and site characteristics.

If the existing information is not sufficient to adequately characterize the waste, then field sampling of the wastes will be conducted. The specific sampling media, methods and analytical methods will be SWMU specific. The extent of this characterization will be sufficient to select sampling locations and the analytical methods for sampling adjacent soils, surface water or groundwater, whichever is the first potentially affected media.

Release Determination

Sampling of underlying or adjacent soils will be the primary method for determining if a release has occurred. Sampling locations and analytical methods will be consistent with the SWMU waste characteristics and SWMU construction. Sampling locations will be selected to ensure that all likely migration routes from the SWMU (such as pipeline leakage, liner leakage, etc.) are investigated.

Sampling will also be conducted to establish SWMU background concentrations for the constituents of interest. These concentrations will provide the basis for determining whether any contaminants which are detected are significant enough to indicate a release from the SWMU.

Depending upon site and waste characteristics, other investigation methods such as soil vapor surveys, groundwater monitoring wells, electromagnetic surveys, and pipeline or tank testing methods may be utilized to determine whether a release has occurred.

CLOSURE OF SWMU

If it has been determined that SWMU did release contamination into the environment, a corrective action and/or closure would be initiated. Refer to Chapter 16 for details on the corrective action process. If the facility is no longer in use, it should be cleaned up according to a closure plan. The closure plan would detail the necessary steps to remove waste and decontaminate the SWMU to acceptable limits. Refer to 40 CFR, Subpart G, *§§* 264.110 through 264.120 for detail.

CHAPTER 16

THE CORRECTIVE ACTION PROCESS

In the Hazardous and Solid Waste Amendments to RCRA, Congress expanded EPA's authority to address releases of hazardous waste through "corrective actions" (actions taken by a TSDF to clean up releases caused by facility operations) beyond those contained in 40 CFR Part 264 Subpart F. The new authorities allow EPA to address releases to groundwater and all other environmental media at all solid waste management units at TSDFs. Corrective action requirements are imposed through a permit or an enforcement order. The TSDF owner or operator is responsible for complying with these requirements. Permits issued to RCRA facilities must, at a minimum, contain schedules of compliance to address releases and include provisions for financial assurance to cover the cost of implementing the corrective measures.

SCOPE OF THE CORRECTIVE ACTION PROCESS

To better understand the scope of the corrective action requirements, one must understand its key terms.

Solid waste management units (SWMUs) are waste management units from which hazardous wastes or constituents may migrate, even if the unit was not intended for the management of hazardous waste. Additionally, any areas that become contaminated as a result of routine and systematic releases or wastes are SWMUs (e.g., spill areas).

Regulated units are a subset of all SWMUs. A regulated unit is any surface impoundment, waste pile, land treatment unit, or landfill that received waste after July 26, 1982.

Hazardous constituents are any substances listed in 40 CFR Part 261 Appendix VIII.

The scope of the corrective action process for regulated units at permitted facilities can vary somewhat from that required at other solid waste management units at permitted or interim status facilities. Releases to groundwater from regulated units are addressed under 40 CFR Part 264, Subpart F. Sections 3004(u) and (v) of RCRA (codified in 40 CFR § 264.101) require corrective action for releases of hazardous wastes or constituents from any SWMU at a TSDF that is seeking or subject to an RCRA permit. Section 3004(v) authorizes EPA to impose corrective action requirements for releases that have migrated beyond the facility boundary. Section 3008(h) authorizes EPA to require corrective action or other necessary measures through an enforcement order whenever there is or has been a release of hazardous wastes or constituents from an interim status RCRA facility.

EPA can require permitted facilities with releases from regulated units to:

- Take corrective action only on those releases to the uppermost aquifer (under 40 CFR part 264 Subpart F); or
- Clean up any other contaminated media (under Sections 3004(u) and (v)).

The decision is made by EPA on a case-by-case basis, taking into account the nature and magnitude of the release.

CORRECTIVE ACTION COMPONENTS

The corrective action process has four main components. Each component comprises a number of steps. The number of steps required and the complexity of corrective action permit conditions or other enforcement actions may vary depending on the extent and severity of releases of hazardous wastes at a TSDF. The decision on which steps to include is made on a facility-by-facility basis. EPA also can require that facilities take interim corrective measures whenever necessary to protect human health and the environment.

RCRA Facility Assessment (RFA)

Release determinations for all environmental media from SWMUs (i.e., soil, groundwater, subsurface gas, air, or surface water) will be made by the regulatory agency primarily through the RCRA Facility Assessment (RFA) process. The regulatory agency will perform the RFA for each facility seeking a RCRA permit to determine if there are continuing releases of concern. The major objectives of the RFA are to:

- Identify SWMUs and collect existing information on contaminant releases; and
- Identify releases or suspected releases needing further investigation.

The RFA begins with a preliminary but fairly comprehensive review of pertinent existing information on the facility. If necessary, the review is followed by a visual site inspection to verify information obtained in the preliminary review and to gather information needed to develop a sampling plan. A sampling visit is subsequently performed, if necessary, to obtain appropriate samples for making release determinations.

The findings of the RFA will result in one or more of the following actions:

- No further action under the RCRA corrective action program is required at this time, since no evidence of a release(s) or of a suspected release(s) was identified;
- An RFI by the facility owner or operator is required where the information collected indicates a release(s) or suspected release(s) that warrants further investigation;
- Interim corrective measures by the owner or operator are required where the regulatory agency believes that expedited action should be taken to protect human health or the environment; and
- In cases where problems associated with permitted releases are found, the regulatory agency will refer such releases to the appropriate permitting authorities.

RCRA Facility Investigation (RFI)

If the regulatory agency determines that an RFI is necessary, this investigation will be required of the owner or operator either under a permit schedule of compliance or under an enforcement order. The regulatory agency will apply the appropriate regulatory authority and develop specific conditions in permits or enforcement orders. These conditions will generally be based on results of the RFA and will identify specific units or releases needing further investigation. Such permits or orders may be accompanied by a supporting fact sheet. The RFI can range widely from a small specific activity to a complex multimedia study. In any case, through these conditions, the regulatory agency will direct the owner or operator to investigate releases of concern. The investigation may initially involve verification of a suspected release. If confirmed, further characterization of such releases will be necessary.

The RFI step also includes interpretation by the regulatory agency of release characterization data against established health and environmental criteria to determine whether a Corrective Measures Study (CMS) is necessary. This evaluation is crucial to the corrective action process. The regulatory agency will ensure that data and information collected during the RFI adequately describe the release, and can be used to make decisions regarding the need for a CMS with a high degree of confidence.

Identifying and implementing interim corrective measures may also be conducted during the RFI. If, in the process of conducting the investigation, a condition is identified that indicates that adverse exposure to hazardous constituents is presently occurring or is imminent, interim corrective measures may be needed. Both the owner or operator and the regulatory agency have a continuing responsibility to identify and respond to emergency situations and to define priority situations that warrant interim corrective measures.

Corrective Measures Study (CMS)

If the potential need for corrective measures is identified during the RFI process, the owner or operator is then responsible for performing a Corrective Measures Study (CMS). During this step of the corrective action process, the owner or operator will identify and recommend as appropriate, specific corrective measures that will correct the release.

Information generated during the RFI will be used not only to determine the potential need for corrective measures, but also to aid in the selection and implementation of these measures. While conducting the RFI, the owner or operator is encouraged to collect data (e.g., engineering data such as soil compacting properties or aquifer pumping tests) which may be needed to select and implement corrective measures.

Corrective Measures Implementation (CMI)

Corrective measures implementation includes designing, constructing, operating, maintaining, and monitoring selected corrective measures. If the remedy is not properly implemented, EPA will direct the facility to take additional action on a site-specific basis.

HSWA requires that facilities demonstrate financial assurance for corrective action prior to implementation. This ensures that facilities have the necessary funds available to carry out cleanup of the site. EPA has proposed regulations to require financial assurance for corrective action. Under the proposed rule, acceptable financial mechanisms include trust funds, surety bonds, letters of credit, financial tests, and corporate guarantees. Until finalized, the proposed rule is used as guidance to implement the statutory requirement for financial assurance for corrective action.

THE ENVIRONMENTAL PRIORITIES INITIATIVE

In overseeing the cleanup of hazardous waste sites, EPA is faced with balancing a number of high priority activities. For example, over 2,000 RCRA facilities are likely to require corrective action; a similar number of sites must be addressed under the Superfund program. To ensure that those sites posing the greatest threat are addressed first, EPA developed the Environmental Priorities Initiative (EPI).

The EPI is an integrated RCRA/Superfund screening approach that is used to ensure that the most environmentally significant facilities and sites are given priority for cleanup.

Under the EPI, all RCRA facilities and Superfund sites receive a ranking of their environmental priority. The ranking, already completed for the bulk of sites, is based on the threat each site poses to human health and the environment. The highest priority facilities will next receive a site inspection. The inspection, combined with the earlier corrective action cleanup process, will be used for active RCRA facilities that are identified as high priority in lieu of the Superfund remedial process, in order to conserve Superfund resources and ensure that owners or operators pay for the site remediation.

CHAPTER 17

MIXTURES, EMPTY CONTAINERS, AND RESIDUES

All generators of hazardous waste should be cognizant of the mixture rule. Although it has been appealed and is under review, the mixture rules state that if a listed hazardous waste is mixed with a non-hazardous waste, the entire mixture is hazardous and carries the listed waste code. Do not mix hazardous waste with non-hazardous waste!

The EPA and the DOT have separate definitions of what is considered an empty container. This chapter provides information to permit one to comply with both agencies' requirements.

Residues from containers may or may not be hazardous and regulated. It is incumbent on the generator to determine if they are regulated. Please be aware that improper disposal of residues may expose the generator to liability under CERCLA.

MIXTURES

Mixture Rules

Any mixture of *solid waste* and a *listed* hazardous waste is a listed hazardous waste. Mixtures of *solid waste* and *characteristic* hazardous waste is hazardous only if the mixture exhibits a characteristic.

Mixing

Mixtures created by mixing hazardous waste and nonhazardous material is a form of treatment which requires a permit, except under limited exemptions. For example, 40 CFR § 265.1(c)(13) specifies that the addition of absorbent to waste or waste to absorbent, at the time the waste is first placed in a container, is not a form of treatment requiring a permit. Other exemptions include elementary neutralization, some wastewater treatment or pretreatment, and totally enclosed treatment systems.

Federal law allows "in container treatment" in 90-day accumulation areas without a permit, but state laws vary considerably on the subject.

Mixtures of Hazardous Wastes

There is no restriction on mixing two or more hazardous wastes together so long as they are not incompatible or do not create a reaction which endangers containment. 40 CFR Parts 264 and 265 include appendices (Appendix V in each) which list some potentially incompatible wastes. Note that if one of the wastes is an *acutely* hazardous waste, the entirety becomes an acutely hazardous waste.

Mixtures of more than one hazardous waste should be labeled so as to convey the full hazard which may be associated with them. Generally, this means labeling, manifesting, etc., with all of the EPA waste numbers and hazard codes applicable to the separate wastes.

RCRA & DOT "EMPTY" CONTAINERS

EPA has two special sections on containers and residues contained in containers. These are in 40 CFR §§ 261.7 and 261.33(c).

For hazardous waste that is a compressed gas, the container is empty when the pressure in the container approaches atmospheric.

If a container holding a hazardous waste meets the definition of "empty," as specified in 40 CFR § 261.7(b), then both the remaining residue and the container are no longer subject to hazardous waste regulations.

In 40 CFR § 261.7(b), EPA defines an empty container or liner depending on the type of hazardous waste (hazardous or acutely hazardous) that was contained. A container or liner which previously held a hazardous waste is empty if:

(1) all wastes have been removed that can be removed using common practices for the type of container; and

(2) no more than one inch of residue remains on the bottom; or

(3) no more than three percent by weight of the container capacity remains if the container is smaller than 110 gallons; or no more than 0.3 percent by weight remains if larger than 110 gallons.

Under 40 CFR § 261.7(a), the hazardous residues in a RCRA "empty" container or liner are exempt from regulation.

For material which is an *acutely* hazardous waste, a container or liner is empty if:

(1) the container or liner has been triple rinsed with an appropriate solvent; or

(2) the container or liner has been cleaned by a method shown to achieve equivalent results; or

(3) the container has an inner liner which prevented contact between the container and the liner which has been removed.

Please note that the DOT considers a container to be hazardous until the shipper decides that it no longer needs to be classified as such. According to the DOT, it is up to the shipper to clean and purge a container previously used for hazardous materials.

Always be sure that your empties are RCRA and DOT "empties."

RESIDUES

Rinsate generated by the triple rinse procedure to clean a container which formerly held an acutely hazardous waste is by definition a hazardous waste.

Residues which are to be beneficially used or reused or legitimately recycled or reclaimed are not regulated under RCRA (40 CFR § 261.33(c)). An example is the residue in a container which is being returned for refilling with the same product.

Regardless of whether a container is considered to be a RCRA and DOT "empty," the residue (regulated or not regulated) should not be disposed of carelessly.

Drum disposal and reconditioning facilities have become Superfund sites, and generators have become responsible for remediation of the sites to which they sent empty drums, due to the discharge of the residues to the soil on the site. Thus it is incumbent upon the generator to make sure that the drum reconditioner or disposal facility properly disposes of empty drums that formerly held hazardous or acutely hazardous waste, and properly treats any residues removed from the drums.

CHAPTER 18

WASTE OIL REGULATIONS

The simplest and most cost effective way to dispose of used oil is by recycling. Most cities have used oil dealers which will test, load and transport used oil from bulk or 55-gallon drums. The important point from the RCRA administrative perspective is to make sure that generators segregate the used oil, which is not considered a RCRA hazardous waste, from hazardous waste. Nothing but used oil should be added to used oil accumulation containers. Reputable used oil dealers will test used oil for the presence of halogenated compounds. If detected, they will reject the oil. Oil containing halogenated compounds is considered a "F" listed hazardous waste.

It is the responsibility of the RCRA administrator to select a reputable used oil dealer. Try to find one that refines the oil or markets it for its heat content. Do not engage a used oil dealer unless you have inspected his operation and know where the oil is recycled.

The Used Oil Recycling Act of 1980 (UORA) was added to the RCRA statute (as RCRA Section 3014) to increase safe recycling and reuse of used oil. UORA required EPA to establish standards for used oil which protect public health and the environment without imposing undue regulatory and financial burdens upon the used oil recycling system. Those standards are established in CFR Parts 260, 261, 266, 271, 279 and 280. The following chapter details compliance with UORA and subsequently promulgated standards.

BACKGROUND

RCRA Section 3014 provides EPA with the authority to regulate generators, transporters, processing and re-refining facilities, and burners that handle recycled used oil or used oils that are to be recycled. Section 3014 does not, however, provide the Agency with regulatory authority over used oils that are not recycled. The Agency believes that other RCRA authorities and other EPA and non-EPA regulations adequately control the management of used oils that are not recycled.

DEFINITIONS

The current EPA definition for "used oil" at 40 CFR § 260.10 (drawn from section 1004(36) of RCRA) includes any oil that has been refined from crude oil, or any synthetic oil that has been used and as a result of such use is contaminated by physical or chemical impurities.

Note that the current definition does not include used oil residues or sludges resulting from storage, processing, or re-refining of used oils. EPA believes that the types and concentrations of hazardous constituents in used oil residues and sludges are different from those typically found in used oils, and therefore these residues and sludges warrant separate regulatory consideration. EPA is going to continue to study used oil residues and sludges, as well as all of the residuals from used oil re-refining activities, in the interest of future rule making.

The current EPA definition for "recycled oil" in Section 1004 of RCRA includes used oil being reused for any purpose, including used oil being re-refined or being processed into fuel. EPA believes that Section 3014 also provides authority for establishing management standards that specifically include used oil being stored, collected or otherwise managed prior to recycling.

LISTING AND COMPLIANCE

EPA has determined that recycled or disposed used oil does not have to be listed as a hazardous waste since the used oil management standards under RCRA Section 3014 are adequate. These standards cover used oil generators, transporters, processors and re-refiners, burners, and marketers, and are codified in a new Part 279 of Chapter 40 of the Code of Federal Regulations.

Used oil which does become hazardous through use or adulteration can be controlled through the Toxicity Characteristic (TC) rule, issued in 1990, which controls the disposal of hazardous solid waste.

The Toxicity Characteristic rule provides regulatory limits on lead, benzene, and other contaminants that may be present when used oil is produced. Under current regulations, a used oil handler must determine (through testing or knowledge) that the used oil does not exceed the regulatory limits for TC constituents. Used oil that fails the TC must be disposed according to hazardous waste regulations under subtitle C. Used oil that does not exceed the toxicity characteristic is not a hazardous waste.

EPA has further determined that properly drained used oil filters do not exhibit the toxicity characteristic. Therefore, used oil filters are not considered hazardous and need not be regulated when recycled or disposed. EPA continues to encourage recycling of used oil removed from filters, and recycling of the filters and their components.

USED OIL MANAGEMENT STANDARDS RATIONALE

The management standards promulgated specifically address the practices in managing recycled used oil. These are:

1. *Improper storage*. Management standards have stringent secondary containment and spill cleanup provisions for used oil processors and re-refiners. Also, storage of used oil in unlined surface impoundments (unless only *de minimus* amounts of used oil are present) is banned outright.
2. *Road Oiling*. Management standards ban the use of used oil for road oiling and dust suppression purposes.
3. *Adulteration with hazardous waste*. The addition of hazardous waste, or "adulteration," results in a more toxic mixture that may be spilled, burned or even dumped. The management standards address adulteration in four main ways:

- The "rebuttable presumption" provision of 40 CFR Part 266, Subpart E, which currently applies to used oil burned for energy recovery, has been expanded to cover all used oils, regardless of intended disposition;
- All used oil handlers must label their tanks and containers used to store used oil with the term "used oil" to assist employees in identifying which units are used exclusively for used oil storage and to avoid inadvertent mixing with other wastes; and
- The existing invoice system in 40 CFR Part 266, Subpart E, for used oil fuels has been supplemented with a tracking system consisting of acceptance and delivery records. Tracking of used oil shipments applies to all used oil transporters and processing and re-refining facilities. The tracking system will assist in identifying accountability, should mixing be suspected.

PRESUMPTION OF RECYCLABILITY

EPA presumes recyclability for all used oils. The Agency agrees with public comments that the physical characteristic of the used oil and the used oil recycling market will dictate the conditions for recycling of used oil. However, the Agency has retained the recycling presumption because the presumption simplifies the used oil management system by ensuring that generators and others may comply with one set of standards, the Part 279 standards, regardless of whether the used oil exhibits a hazardous characteristic and regardless of whether the used oil will ultimately be recycled or disposed. In other words, the generator (or any other person handling the oil prior to the person who decides to dispose of the oil) need not decide whether the used oil eventually will be recycled or disposed, and thus need not tailor its management of the oil based upon that decision. Rather, the Part 279 standards apply to all used oils until a person disposes of the used oil, or sends it for disposal. The recycling presumption will not apply once the generator or other person disposes or sends the used oil for disposal.

REGULATORY PROGRAMS AND REQUIREMENTS

EPA's promulgated management standards for recycled used oil cover all segments of the used oil recycling system, and are codified in a new Part 279 of 40 CFR. While generators are the largest segment of this industry, the most stringent standards apply to used oil processors and re-refiners because they handle the largest quantities of used oil. The standards are not expected to cause major economic impacts, but are designed to correct and control certain practices. They prohibit storage in unlined surface impoundments and road oiling (except in states authorized to manage their own hazardous waste programs).

Requirements for Service Stations and Other Generators

A generator is any business which produces used oil through commercial or industrial operations, or that collects it from these operations or private households. Besides vehicle repair shops and service stations, some of the more common examples of used oil generators are military motorpools; taxi, bus, and delivery companies; and shipyards. People who change their own oil (do-it-yourselfers) are not covered, nor are farmers who generate an average of 25 gallons or less of used oil per month in a calendar year. Approximately 700,000 facilities qualify as generators.

Generators must:

- Keep storage tanks and containers in good condition;
- Label storage tanks "used oil";
- Clean up any used oil spills or leaks to the environment; and
- Use a transporter with an EPA identification number when shipping used oil off-site.

Service station dealers that comply with these requirements, that send used oil for recycling, and that accept used oil from individuals are not liable for emergency response costs or damages resulting from threatened or actual releases of used oil from subsequent handling of the oil. EPA believes relief from this particular regulatory burden will encourage more service station dealers to collect used oil, thereby increasing used oil recycling by the consumer sector.

Further storage tank standards are found under the Clean Water Act Spill Prevention Control and Countermeasures (SPCC) requirements, which regulate the storage of materials including used oil, and, for storage in underground tanks, under subtitle I of RCRA (40 CFR Part 280).

Requirements for Processors and Re-refiners

Used oil processors and re-refiners handle and store large quantities of used oil for a wide variety of purposes. Consequently, data suggest that damage from mismanagement of used oil at these facilities is not uncommon, and that stronger controls are necessary. Approximately 300 facilities must comply with these management standards.

Requirements for these facilities include:

- Obtaining an EPA ID number and notifying the Agency of any activities concerning used oil;

- Maintaining storage tanks and containers in good condition, and labeling them "used oil";
- Processing and storing used oil in areas with oil-impervious flooring and secondary containments structures (such as berms, ditches or retaining walls);
- Cleaning up any used oil spills or leaks to the environment;
- Preparing a plan and a schedule for testing used oil products;
- Tracking incoming used oil and out-going recycled oil products;
- Maintaining certain records and biennial reporting;
- Managing used oil processing and re-refining residues safely; and
- Ensuring that the facility is properly closed when recycling operations cease.

REQUIREMENTS FOR TRANSPORTERS AND COLLECTORS OF OFF-SPECIFICATION USED OIL

The transport of used oil is regulated under the Department of Transportation's Hazardous Materials Transportation Act (HMTA). Used oil that meets the criteria for being "combustible" or "flammable" is regulated under DOT requirements for classification, packaging, marking, labeling, shipping papers, placarding, recordkeeping and reporting.

A used oil transporter or collector is any person who transports used oil to another site for recycling. Transfer facilities that are holding areas, such as loading docks and parking and storage areas, must comply with the transporter requirements when used oil shipments are held for more than 24 hours en route to their final destination. Generators who transport less than 55 gallons of their own used oil are exempt from the transporter requirements.

Transporters and collectors must obtain an EPA ID number and notify the Agency of any activities concerning used oil; maintain storage tanks and containers in good condition, and label them "used oil"; process and store used oil in areas with oil-impervious flooring and secondary containment structures (such as berms or ditches); clean up any used oil spills or leaks to the environment; and track incoming used oil and out-going used oil. In addition, transporters and collectors must:

- Limit storage at transfer facilities to 35 days; and
- Test waste in storage tanks that are no longer in service for hazardous characteristics, and if wastes are hazardous, close them according to existing hazardous waste management requirements.

Requirements for Used Oil Burners

Used oil burners must comply with the same storage requirements as transporters. Less than 1000 facilities burn off-specification used oil, which must be burned in industrial boilers or furnaces only. The "specification" levels for used oil that will be burned for energy recovery include levels for metals, halogens, and flash point. These existing standards promulgated in 1985 are now recodified from 40 CFR Part 266 to 40 CFR Part 279.

Requirements for Used Oil Marketers

Marketers of used oil were regulated in 1985. These standards are recodified from 40 CFR Part 266, Subpart E to 40 CFR Part 279. There are no major changes to existing requirements, such as those concerning notification, analysis, recordkeeping, and invoices for each shipment.

PCBs and Other Contaminants

The manufacture, use, import, and disposal of oils containing more than 50 ppm of polychlorinated biphenyls (PCBs) are controlled under the Toxic Substances Control Act (TSCA). TSCA also requires reporting of any spill of material containing 50 ppm or greater PCBs, into sewers, drinking water, surface water, grazing lands, or vegetable gardens. CERCLA requires reporting of any 1-pound spill of PCBs into the environment. Note that used oils containing less than 50 ppm of PCBs are covered under RCRA.

Used oils that are contaminated with CERCLA hazardous substances (e.g., due to the presence of elevated levels of lead) are subject to CERCLA release reporting requirements. Therefore, releases of used oil containing such contaminants (e.g., lead) into the environment in quantities greater than the reportable quantity for the contaminant must be reported to the National Response Center. The current RQs for CERCLA hazardous substances are listed in 40 CFR § 302.4. In addition, under 40 CFR Part 110, any discharge of oil that violates applicable water quality standards or causes a film or sheen on a water surface must be reported to the National Response Center.

CHAPTER 19

TOXIC SUBSTANCES CONTROL ACT (TSCA) REGULATED SUBSTANCES

EPA regulates asbestos and polychlorinated biphenyls under authority of the Toxic Substance Control Act. In the past, asbestos was contained in many automotive, consumer, and building products. If friable asbestos is removed from articles or buildings, the waste becomes regulated. Asbestos is an airborne hazard and is contained during removal by isolating the area affected and by wetting. The RCRA administrator's major concern is that asbestos be disposed of properly. Asbestos waste should be double bagged in 6-mil. plastic bags. The bags may be placed in drums or cardboard boxes for shipment. All containers holding asbestos waste must be clearly labeled. There is no requirement for the use of a manifest when shipping asbestos waste, but the use of a chain-of-custody form is recommended. Asbestos waste should be disposed of in a Class I industrial landfill. Information in this chapter is from EPA's "Asbestos Waste Management Guidance" manual of May, 1985, where more exhaustive details can be found.

Polychlorinated biphenyls (PCBs) were commonly used for dielectric fluids in capacitors and transformers. Askarel or fluids containing more than 50 ppm PCBs should be handled as regulated materials. The treatment for PCB-containing fluids is incineration. The treatment for PCB-containing transformers and capacitors is to drain, clean and landfill them. The best treatment is to complete destruction by incineration and smelting and recycling of the metal parts.

TSCA authorizes testing and regulation of chemical substances entering the environment. TSCA supplements sections of the Clean Air Act, the Clean Water Act, and the Occupational Safety and Health Act and is administered by EPA. Compliance with TSCA is directed at management of polychlorinated biphenyls (PCBs) and asbestos and chlorofluoro compounds (Freons).

ASBESTOS WASTE MANAGEMENT

Description of Asbestos

Asbestos is a naturally occurring family of fibrous mineral substance. When disturbed, asbestos fibers may become suspended in the air for many hours, thus increasing the extent of asbestos exposure for individuals within the area.

EPA regulations identify the following types of asbestos: chrysolite, amosite, crocidolite, anthophyllite, actinolite, and tremolite. Approximately 95 percent of all asbestos used in commercial products is chrysolite. Since the early 1900s, asbestos fibers have been mixed with various types of binding materials to create an estimated 3000 different commercial products. Asbestos has been used in brake linings, floor tile, sealants, plastics, cement pipe, cement sheet, paper products, textile products, and insulation. The amount of asbestos contained in these products varies significantly, from 1 to 100 percent, depending on the particular use.

The potential of an asbestos-containing product to release fibers is dependent upon its degree of friability. Friable means that the material can be crumbled with hand pressure and, therefore, is likely to emit fibers. The fibrous or fluffy spray-applied asbestos materials found in many buildings for fireproofing, insulating, soundproofing, or decorative purposes are generally considered friable. Pipe and boiler wrap are also friable and found in numerous buildings. Some materials, such as vinyl-asbestos floor tile, are considered nonfriable and generally do not emit airborne fibers unless subjected to sanding or sawing operations. Other materials, such as asbestos cement sheet and pipe, can emit asbestos fibers if the materials are subjected to breakage or crushing in the demolition of structures that contain such materials. For this reason, such materials are considered friable under the National Emission Standards for Hazardous Air Pollutants (NESHAP) regulations for the demolition of structures.

Identifying Asbestos

Only on rare occasions can the asbestos content in a product be determined from product labeling or by consulting the manufacturer, since most products as placed in use are no longer labeled. A description of common asbestos-containing products is found in Table 1. Further information on asbestos content of consumer products is available through the Consumer Product Safety Commission Hotline:

Continental United States	1/800/638-2772
Maryland only	1/800/492-8363
Alaska, Hawaii, Puerto Rico, Virgin Islands	1/800/638-8333

Positive identification of asbestos requires laboratory analysis of samples. Standard laboratory analysis using polarized light microscopy (PLM) may cost $30 to $60 per sample. For information on locating a laboratory capable of performing the analysis, contact any of EPA's Regional Asbestos Coordinators or call EPA's assistance line at 1/800/334-8571/ext. 6741.

For additional technical information and to obtain EPA's publication regarding sampling and analysis of asbestos entitled "Guidance for Controlling Friable Asbestos-Containing Materials in Buildings" (EPA 560/5-83-002), contact any of EPA's Regional Asbestos Coordinators or call EPA's TSCA hotline: 1/800/424-9065; or in Washington DC.: 554-1404.

Table 1: Summary of Asbestos-Containing Products

Product	Average percent asbestos	Binder	Dates used
Friction products	50	Various polymers	1910-pres.
Plastic products			
Floor tile and sheet	20	PVC, asphalt	1950-pres.
Coatings and sealants		10 Asphalt	1900-pres.
Rigid plastics		50 Phenolic resin	?-pres.
Cement pipe and sheet	20	Portland cement	1930-pres.
Paper products			
Roofing felt	15	Asphalt	1910-pres.
Gaskets	80	Various polymers	?-present
Corrugated paper pipe wrap	80	Starches, sodium silicate	1910-pres.
Other paper	80	Polymers, starches, silicates	1910-pres.
Textile products	90	Cotton, wool	1910-pres.
Insulating and decorative products			
Sprayed coating	50	Portland cement, silicates organic binders	1935-1978
Trowelled coating	70	Portland cement, silicates	1935-1978
Pre-formed pipe wrap	50	Magnesium carbonate, calcium silicate	1926-1975
Insulation board	30	Silicates	Unknown
Boiler insulation	10	Magnesium carbonate, calcium silicate	1890-1978
Other Uses	<50	Many types	1900-pres.

Federal Regulatory Programs

EPA and OSHA have major responsibility for regulatory control over exposure to asbestos. Emissions of asbestos to the ambient air are controlled under Section 112 of the Clean Air Act, which establishes the National Emission Standards for Hazardous Air Pollutants (NESHAPs). The regulations specify control requirements for most asbestos emissions, including work practices to be followed to minimize the release of asbestos fibers during handling of asbestos waste materials. These regulations do not identify a safe threshold level for airborne asbestos fibers. For additional information about the NESHAPs regulations for asbestos, see 40 CFR Part 61, Subpart M.

The OSHA regulations are established to protect workers handling asbestos or asbestos-containing products. The current OSHA regulations include a maximum workplace airborne asbestos concentration limit of 2 fibers/cc on an 8-hour time weighted average basis, and a ceiling limit of 10 fibers/cc in any 15-minute period. The standard includes requirements for respiratory protection and other safety equipment, and work practices to reduce indoor dust levels. For details regarding OSHA's regulations, see 29 CFR Part 1910.

EPA has implemented a separate regulation under the Toxic Substances Control Act (TSCA) to handle the problem of asbestos construction materials used in schools. This regulation requires that all schools be inspected to determine the presence and quantity of asbestos and that the local community be notified as well as the building posted. Corrective actions, such as asbestos removal or encapsulation, are currently left to the discretion of the school administrators. Technical assistance is given through EPA's TSCA hotline: 1/800/424-9065 (554-1404 in Washington, DC). The specific details of the TSCA program are contained in CFR Part 763, Subpart F.

The Asbestos School Hazard Abatement Act of 1984 (ASHAA) establishes a $600 million grant and loan program to assist financially needy schools with asbestos abatement projects. The program also includes the compilation and distribution of information concerning asbestos, and the establishment of standards for abatement projects and abatement contractors. Additional information can be obtained through the toll-free ASHAA hotline: 1/800/835-6700 (554-1404 in Washington, DC).

Wastes containing asbestos are not hazardous wastes under RCRA. However, because state regulations can be more restrictive than the federal regulations under RCRA, some states may have listed asbestos-containing wastes as hazardous wastes. A list of state hazardous waste agencies is available through the RCRA hotline: 1/800/424-9346 (382-3000 in Washington, DC). Current nonhazardous waste regulations under RCRA pertain to facility siting and general operation of disposal sites (including those that handle asbestos). Details concerning these RCRA requirements are contained in 40 CFR Part 257.

Other federal authorities and agencies controlling asbestos include: the EPA which has set standards under the Clean Water Act for asbestos levels in effluents to navigable waters; the Mine Safety and Health Administration, which oversees the safety of workers involved in the mining of asbestos; the Consumer Product Safety Commission; the Food and Drug Administration; and the Department of Transportation.

Removal of Asbestos from Buildings

A significant quantity of asbestos-containing waste may be generated during removal of friable asbestos materials from buildings. EPA regulations address the removal of friable asbestos materials prior to demolition or renovation of buildings in 40 CFR Part 61, Subpart M. Removal should also be considered for materials that may potentially become friable during the demolition or renovation activities. Currently, the federal regulations apply to larger structures; i.e., structures with more than four apartments with certain minimum quantities of asbestos-containing materials. However, some state and local health agencies require removal of lesser quantities of asbestos from smaller buildings.

Regulatory requirements of EPA and OSHA include written advance notice to the regional NESHAPs contact, posting of warning signs, providing workers with protective equipment, wetting friable asbestos material to prevent emissions, monitoring indoor dust levels, and properly disposing of asbestos-containing wastes. It is also highly recommended that the work area be enclosed through the use of plastic barriers to prevent contamination of other parts of the structure. Guidelines for development of an asbestos removal contract are presented in a document entitled "Guide Specifications for the Abatement of Asbestos Releases from Spray- or Trowel-Applied Materials in Buildings and Other Structures," published by the Foundation of the Wall and Ceiling Industry, 25 K Street N.E., Washington, DC 20002 (202/783-6580).

Asbestos removal contractors are encouraged to employ additional safety procedures beyond the minimum requirements of EPA and OSHA. The use of a negative air pressure system, utilizing fans and filters to exhaust air from the room, and a shower decontamination facility for anyone exiting the area are highly recommended. High efficiency particulate air (HEPA) filters, which rate a 99.97 percent removal efficiency for asbestos-size dust, are also recommended. These safeguards better protect workers and prevent contamination of the neighborhood. For additional information, see the EPA document entitled "Guidance for Controlling Friable Asbestos-Containing Materials in Buildings" (EPA 560/5-83-002), available by calling EPA's TSCA hotline: 1/800/424-9065 (554-1404 in Washington, DC).

Waste Handling and Containerization

When the asbestos materials are prepared for removal, they are wetted with a water and surfactant mixture sprayed in a fine mist, allowing time between sprayings for complete penetration of the material. Once the thoroughly wetted asbestos material has been removed from a building component, EPA and OSHA regulations require the wastes to be containerized as necessary to avoid creating dust during transport and disposal. The generally recommended containers are 6-mil thick plastic bags, sealed in such a way as to make them leak-tight. When using plastic bags it is important to minimize the amount of void space or air in the bag. This will help minimize any emissions should the bag burst under pressure. More thorough containerization may include double bagging, plastic-lined cardboard containers, or plastic-lined metal containers. Asbestos waste slurries can be packaged in leak-tight drums if they are too heavy for the plastic bag containers. Both EPA and OSHA specify that the containers be tagged with a warning label. Either the EPA or OSHA label must be used.

CAUTION
CONTAINS ASBESTOS FIBERS
AVOID OPENING OR BREAKING CONTAINER
BREATHING ASBESTOS IS HAZARDOUS TO YOUR HEALTH
or
CAUTION
CONTAINS ASBESTOS FIBERS
AVOID CREATING DUST
MAY CAUSE SERIOUS BODILY HARM

There are currently no regulatory requirements that govern the time period that waste can remain on-site before transport to a disposal site. However, recognizing the health risk and potential liabilities associated with accidental exposure, waste should be guarded (i.e., protected against public access, such as by a fence or in a locked building) and transported as soon as possible.

Cleanup

After the asbestos-containing materials have been removed, all plastic barriers should be removed and the facility should be thoroughly washed. The plastic used to line the walls, floors, etc., should be treated as asbestos waste and containerized appropriately. Cleanup of asbestos debris may be done with a HEPA vacuum cleaner. Any asbestos-containing waste collected by the HEPA vacuum cleaner must be appropriately bagged, labeled and disposed.

All areas of the facility that were potentially exposed to asbestos fibers should be washed down. Several washings should be performed along with air sampling and analysis to assure a low airborne asbestos fiber concentration. Various regulatory agencies have targeted asbestos fiber concentrations in the range of 0.001 to 0.0001 fibers/cc as a level desirable in the building air after cleanup. For example, the State of Arizona has specified 0.001 fibers cc as a level above which additional cleanup is required, and British researchers have identified a level of

0.0001 fibers cc to be attainable after cleanup. In some cases, it may not be possible to remove all asbestos due to the irregularity or porosity of the subsurface materials. In these situations, it may be necessary to spray an encapsulating paint over the surface to eliminate the potential for fiber release. For further information on encapsulants, contact any of the regional Asbestos Coordinators or call EPA's TSCA hotline: 1/800/424-9065 (554-1404 in Washington, DC).

Alternate Handling Techniques

Alternative techniques for removing asbestos materials from buildings must receive prior approval from EPA. To date, the only alternate technique is by vacuum truck. Vacuum trucks will be reviewed by EPA on a case-by-case basis. The one system found to be acceptable by EPA has demonstrated the capability of removing asbestos materials in a wet condition. The asbestos material, contained within the truck as a slurry, is transported to the final disposal site. The air from the vacuum intake is dried and exhausted through a fabric filter located on the truck. Final filtration of exhaust air is through a HEPA filter.

Transport of Asbestos Waste

Transport is defined as all activities from receipt of the containerized asbestos waste at the generation site until it has been unloaded at the disposal site. Current EPA regulations state that there must be no visible emissions to the outside air during waste transport. However, recognizing the potential hazards and subsequent liabilities associated with exposure, the following additional precautions are recommended.

Recordkeeping

Before accepting wastes, a transporter should determine if the waste is properly wetted and containerized. The transporter should then require a chain-of-custody form signed by the generator. A chain-of-custody form may include the name and address of the generator, the name and address of the pickup site, the estimated quantity of asbestos waste, types of containers used, and the destination of the waste. The chain-of-custody form should then be signed over to a disposal site operator to transfer responsibility for the asbestos waste. A copy of the form signed by the disposal site operator should be maintained by the transporter as evidence of receipt at the disposal site.

Waste Handling

A transporter should ensure that the asbestos waste is properly contained in leak-tight containers with appropriate labels, and that the outside of the containers are not contaminated with asbestos debris adhering to the container. If there is reason to believe that the condition of the asbestos waste may allow significant fiber release, the transporter should not accept the waste. Improper containerization of wastes is a violation of the NESHAPs regulation and should be reported to the EPA.

Waste Transport

Although there are no regulatory specifications regarding the transport vehicle, it is recommended that vehicles used for transport of containerized asbestos waste have an enclosed carrying compartment or utilize a canvas covering sufficient to contain the transported waste, prevent damage to containers, and prevent fiber release. Transport of large quantities of asbestos waste is commonly conducted in a 20-cubic yard "roll off" box, which should also be covered. Vehicles that use compactors to reduce waste volume should not be used because these will cause the waste containers to rupture. Vacuum trucks used to transport waste slurry must be inspected to ensure that water is not leaking from the truck.

Disposal of Asbestos Waste

Disposal involves the isolation of asbestos waste material in order to prevent fiber release to air or water. Landfilling is recommended as an environmentally sound isolation method because asbestos fibers are virtually immobile in soil. Other disposal techniques such as incineration or chemical treatment are not feasible due to the unique properties of asbestos. EPA has established asbestos disposal requirements for active and inactive disposal sites under NESHAPs (40 CFR Part 61, Subpart M) and specifies general requirements for solid waste disposal under RCRA (40 CFR Part 257). Advance EPA notification of the intended disposal site is required by NESHAPs.

A landfill approved for receipt of asbestos waste should require notification by the waste hauler that the load contains asbestos. The landfill operator should inspect the loads to verify that asbestos waste is properly contained in leak-tight containers and labeled appropriately. For more detailed disposal procedures, see EPA's "Asbestos Waste Management Guide" manual of May, 1985.

PCB REGULATIONS

On December 21, 1989, the Environmental Protection Agency (EPA) published final regulations amending the disposal and storage requirements for polychlorinated biphenyls (PCBs). The amendments were effective on February 5, 1990, and create notification requirements for certain entities that handle PCB waste, establish a manifest tracking system for PCB waste, add recordkeeping and reporting requirements, and require commercial storers of PCB waste to obtain approval of their storage facilities. This section provides an overview of these new regulations.

Consistent with existing PCB regulations, the new notification and tracking system is promulgated under the Toxic Substances Control Act (TSCA) and parallels closely EPA's manifest tracking system under Subtitle C of the Resource Conservation and Recovery Act (RCRA), 40 CFR Part 262. The new rule does not affect an entity's responsibility to comply with RCRA requirements and does not preempt state laws. Likewise, compliance with RCRA programs requires adherence to both federal and state regulations.

Definition of PCB Waste

PCB waste is defined as those PCBs and PCB items that are subject to the PCB disposal requirements of Subpart D found at 40 CFR § 761.60 *et seq*. Diluting waste to reduce its concentration and therefore avoid disposal requirements is prohibited. PCBs and PCB items do not become subject to the disposal requirements until they no longer serve their intended purpose and are intended for disposal. Intact, non-leaking small capacitors and drained PCB-contaminated equipment are not subject to Subpart D requirements, and thus are not subject to the new rule. Disposal of material resulting from cleanup of a spill of PCBs at a concentration of 50 ppm or greater is included in the new rule even though cleanup residuals may be less than 50 ppm.

Notification of PCB Waste Handling Activities

Persons who store commercially, transport, or dispose of PCB waste, as well as certain generators of PCB wastes, must notify EPA of their waste handling activities no later than April 4, 1990. Only generators who own or operate PCB storage facilities which are subject to 40 CFR § 761.65(b) or (c)(7) must provide the notification. Upon notification, EPA will issue an identification number unless one has been previously issued under RCRA authority. The identification number must then be entered on manifests which accompany PCB waste shipments from the point of generation to the disposal facility. Generators of PCB waste who do not maintain storage areas subject to 40 CFR § 761.65 do not need to notify EPA, and will be deemed to have received by rule the identification number "40 CFR Part 761." After June 4, 1990, PCB waste may not be processed by a generator or accepted for transport, storage, or disposal by or from an entity without an EPA identification number.

Notification is made by filing EPA Form 7710-53 with EPA. Generators with more than one regulated storage facility must provide the notification and obtain an EPA identification number for each facility. Those entities that have previously notified EPA or a state of hazardous waste activities under RCRA and who have received a RCRA identification number must still notify EPA of their current RCRA number on the notification form.

Manifest Tracking System

The Uniform Hazardous Waste Manifest, EPA Form 8700-22, is to be used for PCB waste and all generators of PCB waste at concentrations of 50 ppm or greater must manifest their waste. A "generator of PCB waste" is the person whose decision causes a PCB material still under his physical control to become subject to the PCB disposal requirements (40 CFR § 761.3). Special handling instructions for PCB manifests require the identification of the date of removal from service for disposal. The "special handling instructions" section should be used for entering the PCB waste code number (PCB1 for PCB articles, transformers, capacitors, etc. or PCB2 for PCB containers), and the earliest date of removal from service for disposal of PCB articles, PCB containers, and PCB article containers contained in the shipment. This is required because the PCB waste must be finally disposed of within one year from the date it was taken out of service and designated for disposal.

PCB waste must be accompanied by a manifest when a generator gives it to a transporter for delivery to an off-site commercial storage or disposal facility, or places the waste on its own transport vehicle for shipment to a commercial off-site storage or disposal facility. A manifest is not required if the wastes are being transported to a storage or disposal facility owned or operated by the generator. The PCB rule does not contain a small quantity generator exclusion similar to RCRA; all generators of PCB waste at concentrations of 50 ppm or greater, regardless of quantity, must originate a manifest.

A manifest is not necessary for PCB materials that are not subject to disposal regulations. Therefore, disposal of drained PCB-contaminated carcasses for salvage, decontaminated PCB containers, drained PCB-contaminated containers, and small capacitors do not require manifesting. Also, shipments of empty waste containers do not require manifesting if they are to be used and not disposed. However, at the time of disposal, PCB waste containers, unless decontaminated in accordance with 40 CFR § 761.79, will require a manifest.

The use of the manifest for PCB waste parallels the manifest system for hazardous waste under RCRA. The generator must prepare the manifest to include sufficient copies for the generator, the initial transporter, each intermediate transporter, the designated commercial storage or disposal facility, and a copy that is to be returned to the generator by the designated facility. The generator signs the manifest certification section and obtains the signature and the date of acceptance by the initial transporter; the generator retains one copy and gives the remaining copies to the initial transporter. The transporter then takes the manifest with the PCB waste to the designated facility with each transporter obtaining the signature and date of acceptance of the subsequent transporter or designated facility and keeping one copy. When the waste is delivered to the designated facility, the owner or operator of the designated facility signs and dates each remaining copy to certify receipt of the waste. The last transporter is provided one copy, the designated facility keeps one copy, and the final copy is returned to the generator within 30 days of receipt by the designated facility. The PCB rule requires the generator to confirm with the designated facility the authenticity of the signed returned copy. The generator must confirm receipt of the waste by telephone, or other means of confirmation agreed to by the generator and commercial storer or disposer, by close of business the day after receipt of the hand-signed manifest.

Reporting, Recordkeeping, and Records

The generator must keep its initial copy of the manifest until it receives the signed copy from the commercial storer or disposal facility. This signed copy must be retained for at least three years from the date the waste was accepted by the initial transporter. Generators who use or store 45 kilograms (99.4 pounds) of PCBs contained in PCB containers, or one or more PCB transformer, or 50 or more PCB large capacitors, are subject to additional recordkeeping requirements, including an annual documents log which must be prepared by July 1 for the previous calendar year. These generators must also retain signed manifests and all other records for at least three years after they cease using and storing PCBs and PCB items in the quantities described above.

Reports

Exception Reporting: If a generator does not receive a copy of the signed manifest from the designated facility within 35 days of the date the waste was accepted by the initial transporter, the generator must contact the transporter and/or the designated facility to determine whether the waste was received. If the generator does not receive a signed copy of the manifest from the designated facility within 45 days from the date the waste was accepted by the initial transporter, the generator must file an Exception Report with the EPA Regional Administrator for the region in which the generator is located. The Exception Report must include a copy of the manifest for which the generator does not have confirmation of delivery and a cover letter explaining efforts taken to locate the PCB wastes and the result of those efforts.

Certificate of Disposal: Within 30 days of the date of disposal of each shipment of manifested PCB waste, the owner or operator of the disposal facility is required to provide the generator with a Certificate of Disposal. Copies are to be retained by both the generator and the disposal facility in accordance with applicable record keeping requirements.

One-Year Exception Reporting: Under existing regulations, PCB waste must be disposed of within one year from the date the PCBs or PCB item is taken out of service and designated for disposal. Generators are presumed to be in compliance with this one-year limit on storage if they can demonstrate that the storage period prior to delivery to the disposal facility did not exceed nine months. As noted above, the generator must indicate on the manifest the date the PCB material was removed from service for disposal. This date notifies subsequent waste handlers of the time by which lawful disposal must occur.

A One-Year Exception Report is required under two circumstances. First, disposal facilities are to submit such reports when they receive PCB wastes on a date more than nine months after the date of removal from service if, because of other disposal commitments, they cannot dispose of the waste within one year from the date of removal from service. Second, generators are required to submit a One-Year Exception Report when they transfer PCB waste to a disposal facility more than nine months from the date of removal from service and either (i) do not receive a Certificate of Disposal confirming the disposal of the PCB waste within 13 months of the date of removal from service or (ii) receive a Certificate of Disposal with a confirmed date of disposal more than one year from the date of removal from service.

Discrepancy Reporting: Discrepancies in the weight, count and type of waste are to be recorded on the manifest. The owner or operator of the designated facility must attempt to reconcile discrepancies with the generator or transporter(s). If the discrepancy cannot be resolved within 15 days after receipt of the waste, the owner or operator must immediately submit a letter describing the discrepancy and its attempts to reconcile the discrepancy, with a copy of the relevant manifest, to the Regional Administrator in the region where the designated facility is located.

Annual Documents and Reports: Existing PCB regulations impose annual document preparation requirements on facilities that use and store their own PCBs or PCB items and on commercial storage and disposal facilities. Entities that use and store their own PCBs must summarize for each calendar year the total amounts of PCBs in use or designated for disposal, as well as information about where and when PCB waste was shipped.

The new PCB rule requires facilities that use and store their own PCBs and PCB items to retain records and annual document logs, but does not require the submission of an annual report. Records include manifests, certificates of disposal, and other documents prepared under the manifest tracking system. Then the annual document log must include the EPA identification number of the facility, manifest numbers for every manifest generated, and detailed information regarding PCB wastes that are stored.

Approvals for Commercial Storage Facilities

Pursuant to the new rule, all commercial storers of PCB waste have interim approval to operate until August 2, 1990 after which date such facilities are prohibited from storing any PCB waste unless they have submitted a complete application for final storage approval. If a complete application is submitted by August 2, 1990, interim approval continues until the final decision on the storage application.

A "commercial storer" of PCB waste means the owner or operator of a storage facility which is subject to the storage facility standards under 40 CFR § 761.65 and which engages in storage activities involving PCB waste generated by others, or PCB waste that was removed while servicing the equipment owned by others and brokered for disposal. The receipt of a fee or other form of compensation for storage services is not necessary to qualify as a commercial storer of PCB waste. A generator who stores only its own waste is subject to PCB storage standards, but is not required to seek approval as a commercial storer. Common storage between related companies (parent and subsidiaries, sibling companies, and companies owned by a common holding company) does not create commercial storage status. Common storage between unrelated companies is considered to be commercial storage.

Capacitors

After October 1, 1988, use of PCB large capacitors is prohibited unless the capacitor is used within a restricted-access electrical substation or in a contained and restricted-access indoor installation. Large capacitors are those that contain more than 1.35 kilograms or 3 pounds of dielectric fluid.

Marking Requirements

All PCB containers, transformers and capacitors must be marked with the PCB marking mark as well as each storage area used to store any PCB item. As of December 1, 1985 the vault door, machinery room door, fence, hallway, or other means of access to a PCB transformer must be marked with the PCB marking.

(We would like to thank Holland and Hart Attorneys at Law for permission to reprint sections of Polychlorinated Biphenyls (PBS) Regulations, dated July, 1990.)

CHLOROFLUOROCARBONS (CFCS)

On March 29, 1990, EPA promulgated revisions to the Toxicity Characteristic (TC), adding 25 new organic constituents. Chloroform and carbon tetrachloride ($CC1_4$) are among the constituents. These constituents are present in some unused chlorofluorocarbon (CFC) refrigerants, and will remain in the refrigerants throughout their use. In certain cases, chloroform and carbon tetrachloride may be present at concentrations that will cause the used refrigerants to exhibit the TC. However, EPA is concerned that regulating used refrigerants as hazardous wastes may create disincentives to reclaiming refrigerants and may increase the venting of used refrigerant gases during maintenance and repair work. Therefore, EPA has suspended the hazardous waste regulations' coverage of used CFC refrigerants if they are reclaimed for reuse. The exemption does not apply to refrigerants that are collected for disposal.

UNDERGROUND STORAGE TANKS

Underground storage tanks are not directly the responsibility of the RCRA administrator but because of the practice and use of tanks to store hazardous substances, or RCRA hazardous waste, this section is included to provide the basics of underground storage tank regulations.

Most of the information in this chapter is from EPA's *Musts for USTs* (EPA/530/UST-88/008, July 1990).

INTRODUCTION

Several million underground storage tank systems in the United States contain petroleum or hazardous chemicals. Tens of thousands of these USTs, including their piping, are currently leaking. It is better to determine if you have a leaker than to be reported for groundwater contamination. Check your tanks.

REGULATORY DEFINITION OF UNDERGROUND STORAGE TANK

An UST is any tank, including underground piping connected to the tank, that has at least 10 percent of its volume underground. The regulations apply only to USTs storing either petroleum or certain hazardous chemicals. Special requirements for chemical USTs are at the end of this chapter. Generally, the requirements for both petroleum and chemical USTs are very similar.

Some kinds of tanks are not covered by these regulations:

- Farm and residential tanks holding 1100 gallons or less or motor fuel used for noncommercial purposes;
- Tanks storing heating oil used on the premises where it is stored;
- Tanks on or above the floor of underground areas, such as basements or tunnels;
- Septic tanks and systems for collecting storm water and wastewater;
- Flow-through process tanks;
- Tanks holding 110 gallons or less; and
- Emergency spill and overfill tanks.

Other storage areas that might be considered "tanks" are also excluded, such as surface impoundments and pits. Some "tanks," such as field-constructed tanks, have been deferred from most of the regulations. The regulations published in the Federal Register fully identify various tank types and which requirements apply to them.

REGULATIONS FOR USTs INSTALLED AFTER DECEMBER 1988

If you install an UST after December 1988, it must meet the requirements for new USTs concerning correct installation, spill and overfill prevention, corrosion protection, and lead detection.

Installation
First, install USTs correctly by using qualified installers who follow industry codes. Faulty installation is a significant cause of UST failures, particularly piping failures. You must also make sure that the contents you store are compatible with the UST system.

Second, you will need to certify on a notification form (see "Reporting and Recordkeeping" section below) that you have used a qualified installer who can assure you that your UST has been installed correctly.

Preventing Spills and Overfills

Because human error causes most spills and overfills, these mistakes can be avoided by following the correct tank filling practices required by the UST regulations. If you and your distributor follow these practices, nearly all spills and overfills can be prevented. Also, the UST regulations require the use of mechanical devices, such as spill catchment basins and overfill alarms, to prevent these releases from harming the environment.

Protecting Tanks and Piping from Corrosion

Tanks and piping must be protected or they will be eaten away by corrosion:

- Steel tanks and piping can be coated with a corrosion-resistant coating and "cathodically" protected. (Cathodic protection uses either sacrificial anodes or impressed current.)
- Tanks and piping can be made totally of a noncorrodible material, such as fiberglass-reinforced plastic. (Metal piping connected to noncorrodible tanks still requires corrosion protection.)
- Tanks and piping can be protected by other methods approved by the regulatory authority.
- Steel tanks (but not piping) can be protected using a method in which a thick layer of noncorrodible material is bonded to the tank.

Detection of Leaks from Tanks

You must check your tanks at least once a month to see if they are leaking, and must use one (or a combination) of the following monthly monitoring methods:

- Automatic tank gauging;
- Monitoring for vapors in the soil;
- Interstitial monitoring; and
- Other approved methods.

For young tanks, you have an additional leak detection choice, but only for 10 years after you install your UST. Instead of using one of the monthly monitoring methods noted above, you can check for leaks by combining monthly inventory control with tank tightness testing every 5 years. After 10 years, you must use one of the monthly monitoring methods listed above.

Some small tanks may be able to use manual tank gauging as a leak detection method, either by itself or in combination with tank tightness testing.

Detection of Leaks from Piping

Because most leaks come from piping, your piping must have leak detection.

If your piping is pressurized, you must meet the following requirements:

- The piping must have devices to automatically shut off or restrict flow or have an alarm that indicates a leak.
- You have to either conduct an annual tightness test of the piping or use one of the following monthly methods noted above for tanks: vapor monitoring, ground-water monitoring, interstitial monitoring, or other approved monthly methods.

If your UST has suction piping, your leak detection requirements will depend on which type of suction piping you have:

- The most commonly used suction piping requires either monthly monitoring (using one of the four monthly methods noted above for use on pressurized piping) or tightness testing of the piping every three years.
- Another kind of suction piping is safer and does not require leak detection. This safer method has two main characteristics:

  (a) Below-grade piping is sloped so that the piping's contents will drain back into the storage tank once the suction is released.

  (b) Only one check valve is included in each suction line and is located directly below the suction pump.

REGULATIONS FOR EXISTING USTs

Existing USTs are those installed before December 1988. In addition to immediately starting tank filling procedures that will prevent spills and overfills, you must meet two major requirements:

- Requirements for corrosion protection and spill and overfill prevention.
- Leak detection requirements.

Requirements for Corrosion Protection and Spill and Overfill Prevention

By December 1998 (10 years after the UST regulations become effective), USTs that were installed before December 1988 must have:

- Corrosion protection for steel tanks (see 22.3.3).
- Devices that prevent spills and overfills (see 22.3.2).

Although the regulatory deadline is in 1998, you should make these improvements as soon as possible to reduce the chance that you will be liable for damages caused by releases from substandard USTs.

Deadlines and Choices for Leak Detection

Leak detection deadlines are being phased in for existing USTs depending on their age:

If the tank was installed . . .	It must have tank detection by December of . . .
before 1965 or unknown	1989
1965-1969	1990
1970-1974	1991
1975-1979	1992
1980-12/1988	1993

This schedule assures that the older USTs, which are more likely to leak, have leak detection first.

You have three basic choices for making sure your tanks are checked at least monthly to see is they are leaking:

- You can use any of the monthly monitoring methods listed for new tanks. Also, very small tanks may be able to use manual tank gauging.
- If your UST has corrosion protection or internal tank lining and devices that prevent spills and overfills, you can combine monthly inventory control with tank tightness testing every 5 years. This choice, however, can only be used for 10 years after adding corrosion protection or internally lining the tank (or until December 1998, whichever date is later). After 10 years, you must use one of the monthly monitoring methods listed above.
- If your UST does not have corrosion protection or internal tank lining and devices that prevent spills and overfills, you can combine monthly inventory control with annual tank tightness testing. Please note, however, that this method is allowed only until December 1998. After that, your UST (now equipped with corrosion protection or an internal tank lining, and devices that prevent spills and overfills) must use one of the first two leak detection choices noted above.

Some choices may be better than others. You have a leak detection advantage if your UST has been "upgraded" with corrosion protection and devices to prevent spills and overfills. For 10 years after "upgrading," you can use a leak detection method that will be less costly and easier to apply than most other leak detection methods. This method requires you to conduct monthly inventory control and to have tank tightness tests performed every 5 years. By contrast, USTs that have not been upgraded must have tank tightness tests every year.

For existing piping, you have two basic choices of leak detection depending on the type of piping you use:

- By December 1990, existing pressurized piping must meet the leak detection requirements for new pressurized piping.
- Existing suction piping must meet the requirements for new suction piping at the same time the tank meets the leak detection schedule given above.

CORRECTIVE ACTIONS

Various warning signals indicate that your UST may be leaking and creating problems for the environment and your business. You can avoid most of these problems by paying careful attention to these warning signals and by taking the appropriate actions.

Suspected Leaks

Look for warning signals of leaks in your equipment and in the environment. The following warning signals may be detected in your equipment:

- Unusual operating conditions (such as erratic behavior of the dispensing pump).
- Results from leak detection monitoring and testing that indicate a leak.

You need to confirm quickly whether these suspected leaks are real. What at first appears to be a leak may be the result of faulty equipment that is part of your UST system or its leak detection. Double check this equipment for failures. You may simply need to repair or replace equipment that is not working.

If repair or replacement of faulty equipment does not solve the problem, then you must report this finding to the regulatory authority and conduct tightness tests of the entire UST system. If these tests indicate a leak, you need to report to the regulatory authority and follow the actions for a confirmed leak.

The environment at or near your site may also suggest evidence of a leak. For example, neighbors might tell you they have smelled petroleum vapors in their basements or tasted petroleum in their drinking water. You might even discover evidence of environmental damage as you investigate the suspected equipment failures discussed above.

Whenever evidence of environmental damage is discovered, you must take the following actions:

- Report this discovery immediately to the regulatory authority.
- Conduct tightness tests of the entire UST system.
- Investigate the UST site for additional information on the extent and nature of the environmental damage.

If a leak is indicated by these tests and site checks, then you will need to follow the actions for responding to confirmed leaks.

Confirmed Leaks

Your response to confirmed leaks and spills (including overfills) comes in two stages -- short-term actions and long-term actions.

Short-term actions:

- Take immediate action to stop and contain the leak or spill.
- Tell the regulatory authority within 24 hours that there is a leak or spill. However, petroleum spills and overfills of less than 25 gallons do not have to be reported if you immediately contain and clean up these releases.
- Make sure the leak or spill poses no immediate hazard to human health and safety by removing explosive vapors and fire hazards. Your fire department should be able to help or advise you with this task. You must also make sure you handle contaminated soil properly so that it poses no hazard (for example, from vapors or direct contact).
- Find out how far the petroleum has moved and begin to recover the leaked petroleum (such as product floating on the water table).
- Report your progress and any information you have collected to the regulatory authority no later than 20 days after you have confirmed a leak or spill.
- Investigate to determine if the leak has damaged or might damage the environment. You must report to the regulatory authority what you have learned from a full investigation of your site within 45 days of confirming a leak or spill. At the same time, you must also submit a report explaining how you plan to remove the leaked petroleum, if you have found contaminated groundwater. Additional site studies may be required if necessary.

Some leaks and spills will require additional, long-term attention to correct the problem.

Long-term actions may include:

- Developing and submitting a Corrective Action Plan that shows how you will meet requirements established for your site by the regulatory authority.

- Making sure you meet the requirements approved by the regulatory authority for your site.

Repairing Tanks and Piping

You can repair a leaking tank if the person who does the repair carefully follows standard industry codes that establish the correct way to conduct repairs.

Within 30 days of the repair, you must prove that the tank repair has worked by doing one of the following:

- Having the tank inspected internally or tightness tested following standard industry codes.
- Using one of the monthly leak detection monitoring methods (except for the method combining inventory control and tank tightness testing).
- Using other methods approved by the regulatory authority.

Within 6 months of repair, USTs with cathodic protection must be tested to show that the cathodic protection is working properly.

You must keep records for each repair as long as you keep the UST in service.

Damaged metal piping cannot be repaired and must be replaced. Loose fittings can simply be tightened, however, if that solves the problem.

Piping made of fiberglass-reinforced plastic, however, can be repaired, but only in accordance with the manufacturer's instructions or national codes of practice. Within 30 days of the repair, piping must be tested in the same ways noted above for testing tank repairs (except for internal inspection).

CLOSURE REQUIREMENTS

You can close your UST permanently or temporarily.

Closing Permanently

If your tank is not protected from corrosion and it remains closed for more than 12 months or you decide to close it permanently, you must follow requirements for permanent closure:

- Notify the regulatory authority 30 days before you close your UST.
- Determine if leaks from your tank have damaged the surrounding environment. If there is damage, you will need to take corrective actions.
- Either remove the UST from the ground or leave it in the ground. In both cases, the tank must be emptied and cleaned by removing all liquids, dangerous vapor levels, and accumulated sludge. These potentially very hazardous actions need to be carried out carefully by following standard safety practices. If you leave your UST in the ground, you

must also fill it with a harmless, chemically inactive solid, like sand. The regulatory authority will help you decide how best to close your UST so that it meets all local requirements for closure.

There are exceptions to these requirements for permanent closure. The requirements may not apply if your UST meets one of the following conditions:

- If your UST meets the requirements for a new or upgraded UST, then it can remain "temporarily" closed indefinitely as long as it meets the requirements below for a temporarily closed UST.
- The regulatory authority can grant an extension beyond the 12-month limit on temporary closure for USTs unprotected from corrosion.
- You can change the contents of your UST to an unregulated substance, such as water. Before you make this change, you must notify the regulatory authority, clean and empty the UST, and determine if any damage to the environment was caused while the UST held regulated substances. If there is damage, then you must take corrective actions.

Closing Temporarily

Tanks not used for 3 to 12 months must follow requirements for temporary closure:

- If your UST has corrosion protection and leak detection, you must continue to operate these protective systems. If a leak is found, you will have to respond just as you would for a leak from an active UST. (If your UST is empty, however, you do not need to maintain leak detection.)
- You must cap all lines, except the vent-line, attached to your UST.

FINANCIAL RESPONSIBILITY

In general, owners or operators of petroleum USTs must be able to demonstrate their ability to pay for damage that could be caused if their tanks leaked. These payments would need to cover the costs of cleaning up a site and compensating other people for bodily injury and property damage caused by your leaking UST. A complete explanation of your financial responsibility requirements appears in the Federal Register (October 26, 1988) and in an EPA brochure, "Dollars and Sense."

REPORTING AND RECORDKEEPING

In general, you will only need to report to the regulatory authority at the beginning and end of your UST system's operating life:

- When you install an UST, you have to fill out a notification form available from your state. This form provides information about your UST, including a certification of correct installation. (You should have already used this form to identify your existing USTs. If you haven't done that yet, be sure you do so now.)

- You must report suspected releases to the regulatory authority.
- You must report confirmed releases to your regulatory authority. You must also report follow-up actions you plan or have taken to correct the damage caused by your UST.
- You must notify the regulatory authority 30 days before you permanently close your UST.

You need to check with your regulatory authority about the particular reporting requirements in your area, including any additional or more stringent requirements than those noted above.

You will also have to keep records that can be provided to an inspector during an on-site visit that prove your facility meets certain requirements. These records must be kept long enough to show your facility's recent compliance status in four major areas:

- You will have to keep records of leak detection performance and upkeep:

  (a) The last year's monitoring results, and the most recent tightness test.

  (b) Copies of performance claims provided by leak detection manufacturers.

  (c) Records of recent maintenance, repair, and calibration of leak detection equipment installed on-site.

- You will have to keep records showing that the last two inspections of your corrosion protection system were carried out by properly trained professionals.
- You must keep records showing that a repaired or upgraded UST system was properly repaired or upgraded.
- For at least 3 years after closing an UST, you must keep records of the site assessment results required for permanent closure. (These results show what impact your UST has had on the surrounding area.)

You should check with your regulatory authority about the particular recordkeeping requirements in your area.

CHEMICAL USTs

Several hundred chemicals were designated as "hazardous" in Section 101(14) of CERCLA. The UST regulations apply to the same hazardous chemicals identified by CERCLA, except for those listed as hazardous wastes. These hazardous wastes are already regulated under Subtitle C of RCRA and are not covered by the UST regulations. (See 40 CFR Parts 260-270 for the hazardous waste regulations.)

Information on the CERCLA hazardous chemicals is available from EPA through the RCRA/CERCLA Hotline at 1-800-424-9346 or (202) 382-3000.

Requirements for New Chemical USTs

New chemical USTs have to meet the same requirements described earlier for new petroleum USTs concerning correct installation, corrosion protection, spill and overfill prevention, corrective action, and closure.

However, they must have secondary containment and interstitial monitoring as described below.

Secondary Containment

All new chemical USTs must have "secondary containment." A single-walled tank is the first or "primary" containment. By enclosing an UST within a second wall, leaks can be contained and detected quickly.

There are several ways to construct secondary containment:

- Placing one tank inside another tank or one pipe inside another pipe (making them double-walled systems).
- Placing the UST system inside a concrete vault.
- Lining the excavation zone around the UST system with a liner that cannot be penetrated by the chemical.

Interstitial Monitoring

The chemical UST must have a leak detection system that can indicate the presence of a leak in the confined space between the first and second walls. Several devices are available to monitor this confined "interstitial" space. The UST regulations describe these various methods and the requirements for their proper use.

You can apply for an exemption, called a "variance," from the requirement for secondary containment and interstitial monitoring. Getting a variance will require a lot of work. You will have to convince your regulatory authority that your alternative leak detection method will work effectively by providing detailed studies of your site, proposed leak detection methods, and available methods for corrective action. Also, some states may not allow variances.

Requirements for Existing Chemical USTs

Existing UST systems are those installed before December 1988. In addition to immediately starting tank filling procedures that prevent spills and overfills, you will need to meet the same requirements as existing petroleum USTs.

You can meet the leak detection requirements for existing chemical USTs in one of the following three ways:

- After December 1998, your UST must meet the same requirements for secondary containment and interstitial monitoring that apply to new chemical USTs.
- After December 1988, a variance can be granted if you meet the same requirements described above for getting a variance for a new chemical UST.
- Until December 1998, you can use any of the leak detection methods, other than interstitial monitoring, but only if the method you choose can effectively detect releases of the hazardous chemical stored in the UST. (Variances are not required in these cases before December 1998.)
- After December 1998, you must either use secondary containment and interstitial monitoring or get a variance.

Hazardous Chemical Leaks

In the event of a leak or spill, you must follow the same short-term and long-term actions described earlier for petroleum leaks and spills -- except for two modified short-term actions:

- You must immediately report hazardous chemical spills or overfills that meet or exceed their "reportable quantities" to the National Response Center at 1-800-424-8802 or (202) 267-2675.

- You must also report hazardous chemical spills or overfills that meet or exceed their "reportable quantities" to the regulatory authority within 24 hours. However, if these spills or overfills are smaller than their "reportable quantities" and are immediately contained and cleaned up, they do not need to be reported.

You can get information on the "reportable quantities" by calling the RCRA/CERCLA Hotline at 1-800-424-9346 or (202) 382-3000.

CHAPTER 21

RECYCLING

They say you save the best for last. Recycling, product substitution and process modification will save facilities significant costs on the disposal of hazardous waste and can reduce reporting and other regulatory requirements. It will not be difficult for the RCRA administrator to justify in dollars and cents the savings generated by initiating the above mentioned processes.

It costs approximately two hundred dollars per 55-gallon drum to dispose of flammable waste (D001). The cost for non-flammable waste will range from $250 to $600 per drum depending on the constituents. Once the RCRA administrator has profiled the various waste streams generated by the facility, it will become evident that product substitutions will pay in the short and long run.

Recycling not only gives the company a good image, it also saves money. Empty containers that held hazardous substances should be reused for future shipments. There are Industrial Materials Exchange Services that can advertise your waste for reuse as a process material for other manufacturers. Every pound of material exchanged, even for a modest sum, will save on disposal costs.

Many times a cost benefit analysis for replacement of equipment that generated hazardous waste did not weigh the cost of waste disposal, reporting requirements and necessary pollution control equipment. The RCRA administrator should be able to provide ballpark figures for these costs to justify the true net present value for current equipment, versus the purchase of new equipment that is less "polluting."

PRACTICAL GUIDANCE FOR RECYCLING

At least annually, the RCRA administrator should review each waste stream to identify the candidates for product substitution. The administrator should use the following guidance to select conditions.

Prioritize waste stream by quantities generated. The larger quantity wastes should be addressed sequentially. The waste where a single constituent causes the waste to be hazardous should be scrutinized most closely. Example: A waste stream is hazardous only because it contains methyl ethyl ketone (MEK). Could isopropyl alcohol or other solvents be substituted?

For heavy-metal bearing waste could the metal be removed and recycled? An example would be silver from photo processing.

If aerosol spray paints are being used, could non-aerosol or other paint system materials be used?

In order to minimize hazardous waste disposal: identify wastes (profiling), evaluate likely candidates by quantity and characteristic, and ask the following questions:

- Can waste be reused in other plant operations?
- Can waste be recycled by the generator onsite or offsite?
- Can product substitutions eliminate the waste?
- Can the waste be sold or exchanged?
- Can the process generating the waste be modified?
- Can an in-line (process) treatment system be installed to treat the waste prior to its becoming a waste?

HISTORY

In enacting the Resource Conservation and Recovery Act (RCRA) of 1976, Congress established a national policy calling for the reduction in hazardous waste generation and for the implementation of proper waste management practices in order to minimize the present or future threat to human health and the environment. RCRA explicitly states that "millions of tons of recoverable materials which could be used are needlessly buried each year," that "the recovery and conservation of such materials can reduce the dependence of the United States on foreign resources and reduce the deficit...," and that "solid waste represents a potential source of fuel that can be converted to energy." In outlining these objectives, Congress has made a commitment to encourage recycling and recovery of materials not only as a partial solution to the solid and hazardous waste problem but also as positive national economic policy.

In the Hazardous Solid Waste Amendments (HSWA) of 1984, Congress reiterated this policy. However, there had been many incidents where recyclable materials had been mismanaged at recycling facilities, which were not regulated under RCRA. Many of these facilities had been abandoned and had become listed as Superfund sites, requiring governmental funding for waste removal and remedial activities. Therefore, Congress set forth a mandate in HSWA for the EPA to further establish regulations for specific recyclable materials and recycling facilities which had the potential to be mismanaged. In order to carry out this Congressional mandate, the EPA redefined a solid waste to include all materials being recycled, provided some exclusions for legitimate recycling activities, and then established rules for specific recyclable materials and recycling facilities which they determined required regulation.

DEFINITION OF A SOLID WASTE

The EPA redefined a solid waste to include all materials which are discarded and that were not excluded under 40 CFR § 261.4 (a). A discarded material includes the remainder of solid wastes not specifically excluded and is defined in § 261.4(a) as any material which is:

- abandoned
- recycled
- considered inherently waste-like.

The term ABANDONED includes any material being disposed of, burned or incinerated, or accumulated, stored or treated (excluding recycling) before or in lieu of being abandoned, as stated in 40 CFR § 261.2(b). This includes most materials which are not products or which no longer can be used as products for their intended use. It excludes recycled materials which are discussed below.

The term RECYCLING includes using in a manner that constitutes disposal, burning for energy recovery, reclaiming or accumulating speculatively.

TYPES OF RECYCLING ACTIVITIES

The EPA then further defines these four types of recycling activities as described below.

Reclaimed

Reclamation involves the regeneration of wastes to make them usable products again or recovery of usable materials from wastes. Materials being reclaimed are classified as solid wastes and are subject to regulation with some exceptions. There is an exception for commercial chemical products, off-specification products, and by-products which would only be classified as hazardous waste because they exhibit a characteristic, which EPA feels can be easily reclaimed without being regulated because they are product with value and there is little potential for mismanagement. In addition, sludges, defined in 40 CFR § 260.10 as wastewater and water supply treatment plant sludges and liquors, and which only exhibit a characteristic, are exempted from the classification as a solid waste when reclaimed because they are generally regulated under wastewater permits.

Burned for energy recovery

This classification includes all materials which are burned to recover energy or used to produce a fuel or otherwise contained in fuels as solid wastes, subject to regulation. It includes materials such as waste oils and other wastes with high BTU values which are burned in boilers (described 40 CFR § 260.10) and industrial furnaces.

Used in a manner constituting disposal

This includes any material which is placed on the land, and which was not manufactured as a product intended to be placed on the land, as solid wastes subject to regulation. This also includes materials which are used to produce products for placement on land (wastes mixed with products), in which case the product becomes a solid waste. This does not include products manufactured for placement on land such as soil stabilizers, commercially available fertilizers and other agricultural products, or asphalt. This rule was established to prevent incidents like Times Beach, where dioxin wastes were mixed with oil and used as a dust suppressant on roads, by regulating the mixtures as a solid waste.

Accumulated Speculatively

A material is a solid waste if it is accumulated before being recycled. However, if a person can demonstrate that the material is recyclable and that there is technology available to recycle the material and that at least 75% of the material is recycled or shipped off-site for recycling, then it will not be classified as speculative accumulation. Only pure or technical grade commercial chemical products are exempted from the classification of solid waste when accumulated. This requirement was established to prevent accumulation of waste materials for which the technology for reclamation does not currently exist.

EXEMPTIONS FOR RECYCLING ACTIVITIES

After including recycling activities under the solid waste regulations, as discussed above, the EPA set forth exclusions for those recycling activities which had not historically resulted in the mismanagement of materials. Materials are not solid waste when recycled in the following ways.

- Materials which are used or reused as ingredients in an industrial process without being reclaimed, such as direct use of distillate bottoms from carbon tetrachloride production as a feedstock for carbon tetrachloride production.

- Materials which are used or reused as effective substitutes of commercial products, such as use of waste ammonia as a wastewater treatment ingredient.

- Secondary materials that are returned to the original process (in place of raw materials) from which they were generated prior to a reclamation as in "closed loop" recycling. For example emission control dust recaptured during metal smelting can be returned by burning reclaimed accumulate.

- Air emission-control dust recaptured during metal smelting that can be returned to the smelting process without additional processing and is not considered a solid waste.

REQUIRED DOCUMENTATION

EPA requires that anyone who claims that a material is exempted from solid waste classification must be able to provide documentation. The claimant must be able to show the amount of material stored and recycled, and must substantiate the method of use or reuse.

In addition, recycling facilities must document the following:

- the equipment and technology is available to recycle; and

- that 75% of the material accumulated is recycled annually.

FURTHER EXEMPTIONS FOR RECYCLABLE MATERIALS

The following materials are defined as solid wastes, but are excluded from the hazardous waste requirement as stated in 261.6(a)(3).

- Industrial ethyl alcohol that is reclaimed;

- Used batteries (or used battery cells) returned to a battery manufacturer for regeneration;

- Used oil that exhibits one or more of the characteristics of hazardous wastes but is recycled in some manner other than being burned for energy recovery;

- Scrap metal;

- Fuels produced from the refining of oil-bearing hazardous wastes along with normal process streams at a petroleum refining facility if such wastes result from normal petroleum refining, production, and transportation practices;

- Oil reclaimed from hazardous waste resulting from normal petroleum refining, production, and transportation practices, if it is to be refitted along with normal process streams at a petroleum refining facility;

- Coke and coal tar from the iron and steel industry that contains hazardous waste from the iron and steel production process;

- Hazardous waste fuels which are defined in Section 261.6 are regulated under Part 266.

- Petroleum coke produced from petroleum refinery hazardous waste containing oil at the same facility at which such wastes were generated, unless the resulting coke product exhibits one or more characteristics of hazardous waste.

References

Colorado Hazardous Waste Regulations. Colorado Department of Health, August 1992.

Environmental Guidance Program Reference Book, Comprehensive Environmental Response, Compensation, and Liability Act. Environmental Sciences Division, Oak Ridge National Laboratory, by Martin Marietta Energy Systems, Inc., for Assistant Secretary for Environment, Safety and Health, Office of Environmental Guidance, Revision 11, 15 March, 1991.

Environmental Guidance Program Reference Book, Hazardous Materials Transportation Act. Environmental Sciences Division, Oak Ridge National Laboratory, by Martin Marietta Energy Systems, Inc., for Assistant Secretary for Environment, Safety and Health, Office of Environmental Guidance, Revision 11, 15 March, 1991.

Environmental Guidance Program Reference Book, Resource Conservation and Recovery Act. Environmental Sciences Division, Oak Ridge National Laboratory, by Martin Marietta Energy Systems, Inc., for Assistant Secretary for Environment, Safety and Health, Office of Environmental Guidance, Revision 5, 15 October, 1991.

Holland and Hart, *Polychlorinated Biphenyls (PCB) Regulations,* July, 1990.

J.J. Keller and Associates, Inc. *Hazardous Waste Guide For Generators, Transporters and TSD's,* 1992.

McCoy and Associates, Inc. *The RCRA Land Disposal Rectrictions-- A Guide to Compliance, 1990 Edition.* McCoy and Associates, Inc., Lakewood, Colorado, 1990.

RCRA Orientation Manual, 1990 Edition. U.S. Environmental Protection Agency Office of Solid Waste/Permits and State Programs Division, Office of Ombudsman, and The Association of State and Territorial Solid Waste Management Officials, 1990.

Schleifer, Jay, et al., *How to Comply With Hazardous Waste Laws, A Step-by-Step Guide*, Edition No. 3. Elgin, Illinois S-K Publishing 1991.

Managing Industrial Solid Wastes From Manufacturing, Mining, Oil and Gas Production, and Utility Coal Combustion -- Background Paper. U.S. Congress, Office of Technology Assessment, OTA-BP-O-82. Washington, DC: U.S. Government Printing Office, February 1992.

Transportation of Hazardous Wastes and Substances Training Manual. U.S. Department of Energy, August, 1988.

1990 Emergency Response Guidebook. U.S. Department of Transportation, Research and Special Programs Administration, Office of Hazardous Materials Transportation, 1990.

Hopcroft, Francis J., David L. Vitale, and Donald L. Anglehart. *Hazardous Materials and Hazardous Waste.* Kingston, Massachusetts: R.S. Means Company, Inc., 1989.

Jolley, Robert L., and Rhoda G.M. Wang, eds. *Effective and Safe Waste Management: Interfacing Sciences and Engineering with Monitoring and Risk Analysis.* Boca Raton: Lewis Publishers, 1992.

State Infectious Waste Regulatory Programs. Council of State Governments, 1988.

United States Environmental Protection Agency. *Asbestos Waste Management Guidance*, May, 1985.

United States Environmental Protection Agency. *Environmental News*, September 28, 1992.

United States Environmental Protection Agency. *EPA's Asbestos and Small Business Ombudsman Regulatory Assistance,* February 1, 1992.

United States Environmental Protection Agency. *EPA Guide for Infectious Waste Management*, May, 1986.

United States Environmental Protection Agency. *Federal Register*, 52 (210), October 30, 1987.

United States Environmental Protection Agency. *Federal Register*, 53 (106), June 2, 1988.

United States Environmental Protection Agency. *Federal Register*, 56 (160), August 19, 1991.

United States Environmental Protection Agency. *Federal Register*, 57 (42), March 3, 1992.

United States Environmental Protection Agency. *Federal Register*, 57 (83), April 29, 1992.

United States Environmental Protection Agency. *Federal Register*, 57 (105), June 1, 1992.

United States Environmental Protection Agency. *Federal Register*, 57 (160), August 18, 1992

United States Environmental Protection Agency. *Federal Register*, 57 (208), October 27, 1992.

United States Environmental Protection Agency. *Federal Register*, 57 (211), October 30, 1992.

United States Environmental Protection Agency. *Hazardous Waste Incineration: Questions and Answers*, April 5, 1988.

United States Environmental Protection Agency. *How to Set-Up a Local Program to Recycle Used Oil*, May, 1989.

United States Environmental Protection Agency. *Information for Small Businesses*, March, 1993.

United States Environmental Protection Agency. *Land Disposal Restrictions: Summary of Requirements*, June, 1989.

United States Environmental Protection Agency. *Musts for USTs*, July 1, 1990.

United States Environmental Protection Agency. *Questions and Answers on Land Disposal. Restrictions for Solvents and Dioxins*, May, 1987.

United States Environmental Protection Agency. *Regulatory Assistance for Small Business and Others,* January, 1991.

Wentz, Charles A. *Hazardous Waste Management.* New York: McGraw Hill Book Company, 1989.

APPENDIX A: ACRONYMS

ASHAA	-	Asbestos School Hazard Abatement Act of 1984
ATSDR	-	Agency for Toxic Substances and Disease Registry
BDAT	-	Best Demonstrated Available Technology
BTU	-	British Thermal Unit
CA	-	Corrective Action
CAA	-	Clean Air Act
CASRN	-	Chemical Abstract Substance Registry Number
CCR	-	Colorado Code of Regulations
CDI	-	Case Development Inspection
CEI	-	Compliance Evaluation Inspection
CERCLA	-	Comprehensive Environmental Response, Compensation and Liability Act
CESQG	-	Conditional/ Exempt Small Quantity Generator
CFCs	-	Chloroflourocarbons
CFR	-	Code of Federal Regulations
CHWM	-	Colorado Hazardous Waste Management
CMI	-	Corrective Measures Implementation
CMS	-	Corrective Measures Study
C/PC	-	Closure/Post-closure
CSI	-	Compliance Sampling Inspection
CWA	-	Clean Water Act
DIY	-	Do-It-Yourselfer (people who change their own oil)
DOE	-	Department of Energy
DOJ	-	Department of Justice
DOL	-	Department of Labor
DOT	-	Department of Transportation

EPA	-	Environmental Protection Agency
EPI	-	Environmental Priorities Initiative
ERP	-	Enforcement Response Policy
FFCA	-	Federal Facility Compliance Agreement
FFIS	-	Federal Facility Information System
FID/PID	-	Flame Ionization Detector/ Photoionization Detector
FIFRA	-	Federal Insecticide, Fungicide and Rodenticide Act
FIPS PUB	-	Federal Information Processing Standards Publication
FOIA	-	Freedom of Information Act
FR	-	Federal Register
FY	-	Fiscal Year
GWM	-	Ground-Water Monitoring
GC/MS	-	Gas Chromatograph / Mass Spectrometer
HEPA	-	High Efficiency Particulate Air Filter
HMT	-	Hazardous Materials Table
HMTA	-	Hazardous Materials Transportation Act
HMTR	-	Hazardous Materials Transportation Regulations
HSWA	-	Hazardous and Solid Waste Amendments of 1984
HWMF	-	Hazardous Waste Management Facility
LC50	-	Lethal Concentration for 50% of a population
LD50	-	Lethal Dosage for 50% of a population
LDF	-	Land Disposal Facility
LDR	-	Land Disposal Restrictions
LEPC	-	Local Emergency Planning Committee
LQG	-	Large Quantity Generator
MOA	-	Memorandum of Agreement
MOU	-	Memorandum of Understanding

MQG - Medium Quantity Generator

MSDS - Material Safety Data Sheet

MSWLF - Municipal Solid Waste Landfill

NCP - National Contingency Plan

NESHAP - National Emission Standards for Hazardous Air Pollutants

NOD - Notice of Deficiency

NOV - Notice of Violation

NOS - Not Otherwise Specified

NPL - National Priorities List

NRC - Nuclear Regulatory Commission; National Response Center

NSWMA - National Solid Waste Management Association

NWW - Non-Wastewater

OERR - Office of Emergency and Remedial Response

O&M - Operation and Maintenance

OMB - Office of Management and Budget

O/O - Owner/Operator

ORM - Other Regulated Material - Class

OSH - Occupational Safety and Health

OSHA - Occupational Safety and Health Administration

OSW - Office of Solid Waste

OSWER - Office of Solid Waste and Emergency Response

OUST - Office of Underground Storage Tanks

OWPE - Office of Waste Programs Enforcement

PCBs - Polychlorinated Biphenyls

PLM - Polarized Light Microscopy

PPE - Personal Protective Equipment

RA - Regional Administrator

RCRA	-	Resources Conservation and Recovery Act
RCRIS	-	RCRA Information System
RD&D	-	Research, Demonstration and Development
RFA	-	RCRA Facility Assessment
RFI	-	RCRA Facility Investigation
RQ	-	Reportable Quantity
SAA	-	Satellite Accumulation Area
SARA	-	Superfund Amendments and Reauthorization Act
SERC	-	State Emergency Response Commission
SPCC	-	Clean Water Act Spill Prevention Control and Countermeasures
SQG	-	Small Quantity Generator
SWMU	-	Solid Waste Management Unit
TC	-	Toxicity Characteristic
TCLP	-	Toxicity Characteristic Leaching Procedure
TOC	-	Total Organic Carbon
TSCA	-	Toxic Substances Control Act
TSD	-	Treatment, Storage or Disposal
TSDF	-	Treatment, Storage or Disposal Facility
TSS	-	Total Suspended Solids
TWA	-	Total Waste Analysis
UN / NA	-	United Nations / North American
UORA	-	Used Oil Recycling Act of 1980
UST	-	Underground Storage Tank
VOA	-	Volatile Organics Analysis
VOC	-	Volatile Organic Compounds
WAP	-	Waste Analysis Plan
WW	-	Wastewater

APPENDIX B: ADDITIONAL SOURCES

The field of hazardous waste management is broad and often confusing. The following list should help individual generators and TSDFs to target additional sources for information concerning specific needs.

ASSOCIATIONS

Air & Waste Management Association. Three Gateway Center, Four West, P.O. Box 2861, Pittsburgh, PA 15230; 412/232-3444; FAX: 412/232-3450. 12,000 international professionals from industry, government, consulting and education. Provides forum on technical, scientific, economic, social, political and health related viewpoints on key issues.

Alliance For Responsible Chloroflourocarbon Policy. 1901 North Fort Myer Drive, Suite 1204, Arlington, VA 22209; 703/841-9363. 500 companies that use or produce chloroflourocarbons.

Aluminum Foil Container Manufacturers Association. P.O. Box 1177, Lake Geneva, WI 53147; 800/228-2525; 414/248-9208. Promotes aluminum foil container recyclability and provides a market for same.

Aluminum Recycling Association. 1000 16th Street, N.W., Suite 603, Washington, DC 20036; 202/785-0951; FAX: 202/785-0210. Promotes recycling aluminum base scrap for reuse in automotive, household and industrial products.

American Consulting Engineers Council. 1015 15th Street, NW, Suite 802, Washington, DC 20005; 202/347-7474; FAX: 202/898-0066. Independent, private engineering companies. Provides information on federal legislation, insurance, business practices, international markets and public relations.

American Industrial Hygiene Association. P.O. Box 8390, 345 White Pond Drive, Akron, OH 44320; 216/873-2442; FAX: 216/873-1642. Focus on promoting industrial hygiene through education, training, exchange of information, and Washington representation.

American Society for Testing and Materials. 1916 Race Street, Philadelphia, PA 19103; 215/299-5400. Concerned with keeping 120,000 member civil engineers updated on government regulations and programs. Offers training in recycling, geotechnical sites, and many water programs.

American Society of Mechanical Engineers. 22 Law Drive, Box 2300, Fairfield, NJ 07006-2300; 201/882-1170. Organization of mechanical engineers dealing with all phases of mechanical operations.

American Society of Safety Engineers. 1800 E. Oakton, Des Plaines, IL 60018-2187; 708/692-4121; FAX: 708/296-3769. For safety professionals. Composed of 131 chapters and 47 student sections within 13 regions nationwide. Offers services and professional development through seminars, conferences and newsletter.

Applied BioTreatment Association. P.O. Box 15307, Washington, DC 20003-0307; 202/546-2345; FAX: 202/547-2909. Companies involved in biotreatment and research of oil spills and other environmental contaminants.

Asbestos Information Association of North America. 1745 Jefferson Davis Highway, Arlington, VA 22202; 703/979-1150. The public relations arm of U.S. and Canadian asbestos producers. Provides information on asbestos and health.

Asphalt Recycling & Reclaiming Association. 3 Church Circle, Suite 250, Annapolis, MD 21401; 301/267-0023. Promotes the collective interest of those engaged in the asphalt recycling industry as contractors, manufacturers of equipment, engineers, suppliers and public highway officials.

Asphalt Rubber Producers Group. 3336 North 32nd Street, Suite 106, Phoenix, AZ 85018-6241; 602/955-1141; FAX: 602/956-3506. Promotes the recycling of asphalt/rubber waste products.

Association of Battery Recyclers. P.O. Box 707, Troy, AL 36081; 205/566-1563. Investigates OSHA and EPA compliance methods for the secondary lead smelting industry.

Association of Conservation Engineers. Alabama Dept. of Conservation, 64 N. Union Street, Montgomery, AL 36130; 205/242-3476. Architects and engineers concerned with wildlife conservation and preservation of the environment.

Association of State and Territorial Solid Waste Management Officials. 444 North Capitol Street, NW, Suite 388, Washington, DC 20001; 202/624-5828. State regulatory agents involved with solid waste management engaged in remedial actions under Superfund.

Cement Kiln Recycling Coalition. 1101 30th Street, NW, 5th Floor, Washington, DC 20007; 202/625-3440; FAX: 202/625-3441. Companies that promote safe and beneficial use of waste materials for energy recovery in the cement manufacturing process. Also companies that collect, manage and store wastes for fuel in cement kilns.

Chemical Manufacturers Association. 2501 M Street, N.W., Washington, DC 20037; 202/887-1100. Comprised of 180 chemical manufacturers. Lobbies for environmental, trade, health and safety issues on the federal and state level. Provides legal assistance with litigation.

Chemical Waste Transportation Institute. 1730 Rhode Island Avenue, NW, Suite 1000, Washington, DC 20036; 202/659-4613; FAX: 202/775-5917. Affiliated with National Solid Waste Management Association; Companies must join NSWMA before becoming eligible to join CWTI.

The Chlorine Institute. 2001 L Street NW, Suite 506, Washington DC 20006; 202/775-2790; FAX: 202/223-7225. Producers of gaseous and liquid processing, packaging, transporting, and use.

Coalition for Responsible Waste Incineration. 1330 Connecticut Avenue, NW, Suite 300, Washington, DC 20036; 202/659-0060. Promotes responsible handling and incineration of industrial waste.

Council for Solid Waste Solutions. 1275 K Street NW, Washington, DC 20006; 202/371-5319. A plastics industry group interested in the waste and recycling of their products.

The Environmental Business Association. 1150 Connecticut Avenue, NW, 9th Floor, Washington, DC 20036; 202/862-4363. Promotes the business interests of environmental products manufacturers and service companies.

The Environmental Law Institute. 1616 P Street, NW, Washington, DC 20036; 202/328-5150. National forum for views on environmental issues. ELI publishes journal, deskbook and directory.

Environmental Transportation Association. 122 C Street, NW, Suite 850, Washington, DC 20001; 202/638-7790; FAX: 202/638-1045. National business coalition concerned with the transportation of waste. Promotes development of coordinated plans to solve nation's waste problem.

Hazardous Materials Advisory Council. 1110 Vermont Avenue, NW, Suite 250, Washington, DC 20005-3406; 202/728-1460; FAX: 202/728-1459. Shippers, carriers and container manufacturers involved with hazardous materials.

Hazardous Materials Control Research Institute. 9300 Columbia Boulevard, Silver Spring, MD 20910-1702; 301/587-9390. Professionals involved in the management and disposal of hazardous wastes and/or environmental policy.

Hazardous Waste Action Coalition. 1015 15th Street, NW, Suite 802, Washington, DC 20005; 202/347-7474; FAX: 202/898-0068. An association of engineering and science firms practicing in hazardous waste management.

Hazardous Waste Treatment Council. 1440 New York Avenue, NW, Suite 310, Washington, DC 20005; 202/783-0870. Trade association of waste disposal firms employing high technology treatment techniques rather than land disposal.

Institute of Chemical Waste Management. 1730 Rhode Island Avenue, NW, Suite 1000, Washington, DC 20036; 202/659-4613. Representatives of private sector and municipalities dealing with chemical waste.

Institute of Environmental Sciences. 940 East Northwest Highway, Mount Prospect, IL 60056; 708/255-1561; FAX: 708/255-1699. Engineers, scientists, and educators dedicated to researching, simulating, testing, controlling, and teaching the environments of earth and space.

Institute of Nuclear Materials Management. 60 Revere Drive, Suite 500, Northbrook, IL 60062; 708/480-9573. 1000 members concerned with waste management, physical protection, materials control and accountability.

Institute of Resource Recovery. 1730 Rhode Island Avenue, NW, Suite 1000, Washington, DC 20036; 202/659-4613. Representatives of private sector companies that handle and burn garbage to convert to energy.

Institute of Scrap Recycling Industries. 1627 K Street, Suite 700, Washington, DC 20006; 202/466-4050. Includes processors, brokers and consumers of scrap metal, paper, textiles, glass and plastics.

Institute of Waste Equipment Distributors. 1730 Rhode Island Avenue, NW, Suite 1000, Washington, DC 20036; 202/659-4613. Represents distributors for waste equipment and supplies.

International Association of Environmental Testing Laboratories. 1911 N. Fort Myer Drive, Suite 1105, Arlington, VA 22209; 703/524-2427; FAX: 703/524-1453. Commercial environmental testing laboratories and suppliers to the industry. Examines critical issues pertaining to the testing industry.

International Hazardous Materials Association. 640 East Wilmington Avenue, Salt Lake City, UT 84106; 801/466-3500; FAX: 801/466-9616. Firemen and other first responders. Sponsors annual conference, seminars, quarterly newsletter, chapter meetings.

Manufacturers of Emission Controls Association. 1707 L Street, NW, Suite 570, Washington, DC 20036; 202/296-4797; FAX: 202/331-1388. Manufacturers of automobile exhaust control devices and stationary-source catalytic controls.

National Asbestos Council, Inc. 1777 Northeast Expressway, Suite 150, Atlanta, GA 30329; 404/633-2NAC. Provides comprehensive operations and maintenance training program. Clearinghouse for building owners, environmental professionals, and the public concerning asbestos use.

National Association for Environmental Management. 1440 New York Avenue, NW, Suite 300, Washington, DC 20005; 202/638-1200; FAX: 202/639-8685. Represents environmental managers within corporations and institutions. Primary activities include certification program development, local chapters, liaison with EPA, workshops and seminars.

National Association of Chemical Recyclers. 1333 New Hampshire Avenue, NW, Suite 1100, Washington, DC 20036; 202/463-6956. Companies interested in the advancement of environmentally sound recycling. Member facilities provide spent chemicals for reuse.

National Association of Environmental Professionals. Box 15210, Alexandria, VA 22309-0210; 703/660-2364. Involved in environmental planning, assessment, management, review and research.

National Environmetal Training Association. 8687 Via de Ventura, Suite 214, Scottsdale, AZ 85258; 602/951-1440; FAX: 602/483-0083. Trainers of personnel in the field of air and noise pollution, solid and hazardous waste control, water supply and waste-water treatment.

National Ground Water Association. 6375 Riverside Drive, Dublin, OH 43017; 614/761-1711; FAX: 614/761-3446. Represents the world's ground water industry. Offers short courses and conferences for continuing education; on-line databases.

National Insulation & Abatement Contractors Association. 99 Canal Center Plaza, Suite 222, Alexandria, VA 22314; 703/683-6422; FAX: 703/549-4838. Serves commercial and industrial insulation and asbestos abatement contractors and distributor/fabricators.

National Oil Recyclers Association. Research Oil Company, 277 Broadway Avenue, Cleveland, OH 44115; 216/623-8383. Advances the interests of companies involved in safe recycling of used oil. Members subscribe to a Code of Ethical Business Practices.

National Registry of Environmental Professionals. P.O. Box 2111, Glenview, IL 60025; 708/724-6631; FAX: 708/724-4223. Promotes legal and professional recognition of individuals with education, training and experience as environmental managers, engineers, scientists and technicians.

National Solid Wastes Management Association. 1730 Rhode Island Avenue, NW, Suite 1000, Washington, DC 20036; 202/659-4613; FAX 202/775-5917. Approximately 2500 private, professional firms concerned with collection, transport, management and disposal of solid wastes and refuse.

Plastic Bottle Institute. 1275 K Street, NW, Suite 400, Washington, DC 20005; 202/371-5200. Promotes the application and use of plastic bottles.

Society of the Plastics Industry. 1275 K Street, NW, Suite 400, Washington, DC 20005; 202/371-5200. Promotes application and use of plastics and is the principal representative of the plastics industry.

Spill Control Association of America. 8631 West Jefferson Avenue, Detroit, MI 48209; 313/567-0500; FAX: 313/849-2685. Spill-control contractors, equipment manufacturers, environmental consultants, government officials, training institutions and oil companies. Provides information on legislation, regulation and technologies.

Steel Can Recycling Institute. 680 Anderson Drive, Foster Plaza 10, Pittsburgh PA 15220; 800/876-7274. Provides information and technical analysis to steel companies on methods of collection, preparation, and transportation of steel scrap.

Steel Tank Institute. 570 Oakwood Road, Lake Zurich, IL 60047; 708/438-TANK; FAX: 438-8766. North American companies manufacturing underground and aboveground storage tanks. Conducts research, develops new technologies and standards. Represents members on government issues.

Technical Association of Pulp & Paper Industry. Box 105113, Technology Park, Atlanta, GA 30348; 404/446-1400. Non-profit organization concerned with the pulp and paper industry.

Waste Equipment Manufacturers Institute. A component of the National Solid Wastes Management Association, 1720 Connecticut Avenue NW, Washington, DC 20036; 202/659-4613. Manufacturers of waste handling equipment.

Waste Oil Heating Manufacturers Association. 2550 M Street, NW, Suite 800, Washington, DC 20037; 202/457-6074. Manufacturers of heaters designed to burn waste oil.

Water and Wastewater Equipment Manufacturers Association. Box 17402, Dulles International Airport, Washington, DC 20036; 703/444-1777; FAX: 703/444-1779. Coordinates activities of members in development of sound national water policies and to assure availability of water resource products.

AUDIOVISUAL MATERIALS

ASAP. Dept. HMW, P.O. Box 297, Madison, NJ 07940; 201/377-7308. Hazmat video training. Recent release: *ASAP: Always Suspicious, Always Prepared.*

Bates Video Productions. 1033 O Street, Suite 546, Lincoln, NE 68508; 402/476-7951. Makes and distributes waste management videos.

BLR. 1419 Forest Drive, Annapolis, MD 21403; 800/852-3698; FAX: 301/268-5886. Offers series of five hazardous waste training videos, including: *Employee Introduction to Hazardous Waste Laws; Keeping Track of Hazardous Wastes.*

BNA Communications Inc. 9439 Key West Avenue, Rockville, MD 20850-3396; 800/233-6067. Video training programs for hazardous waste handlers. Recent releases: *Handling Hazardous Waste*, a seven-video series; *Hazardous Waste Training*, complete training program with video, leader's guide and participant handouts.

Comprehensive Loss Management Inc. 7750 West 78th Street, Minneapolis, MN 55439; 800/533-2767. Produces generic and customized video training programs to educate employees and supervisors on chemical hazards. 20 titles with booklets, labeling systems, posters.

Emergency Film Group. 1380 Soldiers Field Road, Boston, MA 02135; 800/842-0999. Emergency response training videos; also has professional personnel available for on-site training.

Environment Video Network. P.O. Box 482, Simcoe, Ontario, CANADA N3Y 4L5; 519/426-3109; FAX: 519/426-0331. Subscribers receive monthly 30-60 minute video program on environmental information and technology.

Federal Publications Inc. 1120 20th Street, NW, Washington, DC 20036; 202/337-7000; FAX: 202/223-0755. Recent title: *A Practical Guide to Federal Environmental Law.* 4 cassette tapes containing 4 hours of major federal environmental laws, policies, regulations and procedures.

Haz-Mat Consultants Inc. P.O. Box 152, West Stockbridge, MA 01266; 413/232-4297. Haz-Mat Film group works with companies to produce a customized training program to keep employees current, as well as legal video documentation of compliance training.

Hazardous Materials Publishing Company. 243 West Main Street, P.O. Box 308, Kutztown, PA 19530; 215/683-6721; FAX: 215/683-3171. Audio-visuals of Transportation Skills Programs on VHS cassettes or 35mm slides with audio cassette based on current DOT/EPA/OSHA hazardous materials, chemicals and waste regulations.

Industrial Training Inc. 5376 52nd Street SE, Grand Rapids, MI 49512; 800/253-4623; 616/698-8688. Offers videos with instruction guides and testing materials. Examples: *Asbestos Awareness; Right-to-Know; Incident Response; Hazardous Waste.*

Institute for Applied Management and Law Inc. 610 Newport Center Drive, Suite 1060, Newport Beach CA 92660; 714/760-1700. Video training products on health and safety laws.

International Video Distributors. 6222 Richmond Avenue, Suite 265, Houston, TX 77057; 713/785-0807. Wide variety of training videos. Examples: *The Environmental Training Series for Oil & Gas Operations; Basic Fire Training for the Petroleum and Chemical Industries; Environmental Compliance, an Employee's Personal Liability; The ABC's of RCRA's Laboratory Training: Basic Lab Skills for the Laboratory Technician.*

J.A.M. Inc. Global Health and Safety Division. 300 Main Street, East Rochester, NY 14445; 716/385-8880; FAX: 716/385-9004. For industries/organizations that require occupational and environmental health and safety training. Recent release: *Biological Safety Training - Infectious Wastes.*

J.J. Keller & Associates Inc. 145 W. Wisconsin Avenue, P.O. Box 368, Neenah, WI 54957-0368; 414/722-2848. Video training programs and software.

NUS Training Corporation. 910 Clopper Road, P.O. Box 6032, Gaithersburg, MD 20877-0962; 800/258-2455; FAX: 301/258-2455. Recent releases: *Bulk Liquid Safety Concerns; Safety Partner; Introduction to HAZ-WOPER; Spill Control Containment; Decontamination; Introduction to HazCom.*

The Marcom Marketing Group Ltd. P.O. Box 9557, Wilmington, DE 19809; 800/654-CHIT; 302/764-3400. Offers training videos. Examples: *Chemical Hazard Identification and Training; The Basic Safety Series; Emergency Preparedness/Crisis Management; Right-to-Know Compliance.*

Training Communications Corp. 483-B Carlisle Drive, Herndon, VA 22070; 703/478-9600; 800/872-2660. Safety Video Training Library, 30 titles available, trainer's guide with each program. Recent releases: *Routes of Chemical Exposure; Environmental Awareness.*

BOOKS AND MANUALS

American Industrial Hygiene Association. 345 White Pond Drive, P. O. Box 8390, Akron, OH 44320; 216/373-2442; FAX: 216/873-1642. Recent titles: *Respiratory Protection; A Strategy For Occupational Exposure.* Contact for full catalog of publications.

American Nuclear Society. 555 N. Kensington Avenue, La Grange Park, IL 60525; 708/352-6611; FAX: 708/352-0499. Recent titles: *The Nuclear Fuel Cycle: Analysis and Management; The Linrar Reactivity Model for Nuclear Fuel Management; Introductory Nuclear Reactor Statics.*

American Petroleum Institute. 1220 L Street, NW, Washington, DC 20005; 202/682-8000. Publishes a wide variety of books, booklets, manuals and training manuals. Send for catalog for complete listing.

American Society of Mechanical Engineers. 22 Law Drive, Box 2300, Fairfield, NJ 07006-2300; 201/882-1170. Recent title: *Hazardous Waste Destruction by High Temperature Incineration.*

American Water Works Association. 6666 West Quincy Avenue, Denver, CO 80235; 303/794-7711 Ext 2615; FAX: 303/794-7310. Publishes more than 400 books. Expanded to include all aspects of the drinking water field and utility management: uses, drinking water treatment and supply, utility operations, microbiology, health effects, risk assessment, engineering practices, youth and consumer education, computer automation, and management.

BLR. 1419 Forest Drive, Annapolis, MD 21403; 800/852-3698; FAX: 301/268-5886. Recent title: *The Book of Chemical Lists: Vol. I and Vol. II.*

Business and Legal Reports Inc. 39 Academy Street, Madison, CT 06443-1513; 800/727-5257; 203/245-7448; FAX: 203/245-2559. Recent titles: *Hazmat, RCRA, Right to Know, Title III; Employee Training Program.*

CRC Press Inc. 2000 Corporate Blvd., NW, Boca Raton, FL33431; 800/272-7737. *Health and Safety in Small Industry: A Practical Guide for Managers.*

Clark Boardman Callahan. 155 Pfingsten Road, Deerfield, IL 60015; 708/948-8955. Publishes books and manuals. Recent offerings: *State and Local Government Solid Waste Management; Environmental Law Series; Clean Air Act Book; Hazard Communication Handbook.*

Federal Publications Inc. 1120 20th Street, NW, Washington, DC 20036; 202/337-7000; FAX: 202/223-0755. Current releases: *Environmental Insurance; Practical Environmental Law; The Environmental Liabilities of Government Contractors & Agencies; The New Clean Air Act.* Also: environmental law manual for over 20 states.

Government Institutes Inc. 966 Hungerford Drive, Suite 24, Rockville, MD 20850; 301/251-9250. Recent titles: *Treatment Technologies: Environmental Statutes; Pretreatment Compliance Inspection and Audit Manual.*

Hazardous Materials Publishing Co. 243 West Main Street, P.O. Box 308, Kutztown, PA 19530; 215/683-6721; FAX: 215/683-3171. Publisher of training manuals and books on codes of federal regulations and many other compliance and training aids. Prints labels and DOT/EPA documentation. Catalog available.

J.J. Keller & Associates Inc. 145 W. Wisconsin Avenue, P.O. Box 368, Neenah, WI 54957-0368; 414/722-2848. Recent offerings: *Chemical Process Industries Catalog*; employee training manuals; material safety data sheets and supplies; hazard communication standard references.

Lewis Publishers Inc. 2000 Corporate Boulevard, NW, Boca Raton, FL 33431-9868; 800/272-7737; 407/994-0555; FAX: 407/997-0949. Recent offerings: *Air Toxics and Risk Assessment; Estimating Costs of Air Pollution Control; Soils in Waste Treatment and Utilization; Biohazards of Drinking Water Treatment; Arthur Young Guide to Water and Waste Water Finance and Pricing.*

Marcel Dekker Inc. 270 Madison Avenue, New York, NY 10016; 212/696-9000. Recent title: *Environmental Management Handbook.* Examines the regulatory and scientific issues of toxic substances management.

McCoy and Associates Inc. 13701 West Jewell Avenue, Suite 202, Lakewood, CO 80228; 303/987-0333; FAX: 303/989-7917. Recent titles: *RCRA Regulations and Keyword Index, 1991 Edition; McCoy's Hazardous Waste Regulatory Update Service; A Guide to Compliance; SARA Title III Regulations and Keyword Index.*

McGraw-Hill Publishing Co. P.O. Box 400, Hightstown, NJ 08520-9403; 609/426-7600. Recent title: *The Standard Handbook of Environmental Engineering.* Provides coverage of current environmental engineering practices, legislation, and criteria to aid in prioritizing work goals.

National Safety Council. 444 North Michigan Avenue, Chicago, IL 60611; 800/621-7619. Publishes over 70 books/manuals on occupational safety and health. Recent titles: *Accident Facts; Work Injury and Illness Rates; Introduction to Occupational Health and Safety.*

OSHA Publications. Room N-3101, U.S. Department of Labor, Washington, DC 20210; 202/523-8148. Recent title: *Air Contaminants: Permissable Exposure Limits.* Newly-issued regulation setting permissable exposure limits for hundreds of toxic and hazardous substances encountered in the workplace. Lists 600 substances and their permissable limits. FREE. Include a self-addressed label.

Pollution Engineering. 1350 E. Touhy Avenue, Des Plaines, IL 60018-3558; 708/635-8800; FAX: 708/390-2636. Publishes booklets and checklists for environmental compliance.

Roytech Publications Inc. 7910 Woodmont Avenue, Suite 902, Bethesda, MD 20814; 301/654-4281; FAX: 301/907-7773. Recent titles: *Chemical Guide to the OSHA Hazard Communication Standard, 6th Edition; Chemical Guide to SARA Title III, 4th Edition.*

Temple University Press. Broad & Oxford Streets, Philadelphia, PA 19122; 215/787-8787. Recent title: *Environmental Accidents, Personal Injury and Public Responsibility.*

The McIlvaine Company. 2970 Maria Avenue, Northbrook, IL 60062; 708/272-0010; FAX 708/272-9673. Recent titles: *(Air) Scrubbers-Adsorbers, FGD and Denox, Fabric Filters; (Water) Liquid Filtration, Sedimentation and Centrifugation; (Energy) Advanced Fuel Technologies.*

The Solid Waste Information Clearinghouse. P.O. Box 7219, Silver Spring, MD 20910; 301/585-2898; 800/67-SWICH; FAX: 301/585-0297. SWICH is the USEP/SWANA national information center for solid waste. Recent publication: The 1991 SWICH Library Catalog. Cost is $40.00.

Van Nostrand Reinhold Company Inc. 115 Fifth Avenue, New York, NY 10003; 606/525-6600. Recent title: *Occupational and Environmental Safety Engineering Management.*

W.H. Brady Co. 727 West Glendale Avenue, P.O. Box 571, Milwaukee, WI 53201-0571; 414/961-2233; 800/635-7557; FAX: 800/445-7446. Recent title: *Hazardous Energy Control Manual.* Includes the lockout/tagout regulations and information for preparing and implementing a plant program to assure compliance. Companion volume: *Hazardous Energy Control Procedures.*

Waste Advantage Inc. 23077 Greenfield Road, Southfield MI 48075; 313/569-8150. Recent title: *Industrial Waste Prevention: Guide to Developing an Effective Waste Minimization Program.* Provides step-by-step instructions for developing a plan to minimize solid and hazardous wastes using an integrated team approach.

Water Environment Federation. Publications Order Department, 601 Wythe Street, Alexandria, VA 22314-1994; 800/556-8700. Recent title: *Standard Methods for the Examination of Water and Wastewater.*

Wiley Law Publications. 7222 Commerce Center Drive, Suite 240, Colorado Springs, CO 80919-9809; 719/548-0200. Recent titles: *Environmental Liability and Real Property Transactions: Law and Practice; Clean Air Act 1990 Amendments: Law and Practice.*

TELEPHONE HOTLINES AND INFORMATION

Air Control Technology Hotline: 919/541-0800. Handles questions concerning the Clean Air Act.

Asbestos Abatement Information Line: 800/368-5888, or in DC and VA, 703/557-1938. Handles questions concerning the Asbestos Emergency Response Act (AHERA) of 1986, and other questions requiring technical response within EPA.

Center For Hazardous Materials Hotline: 800/334-2467. Run by the Penn. Dept. of Human Resources to provide regulatory and technical assistance to business and government.

CHEMTREC Center for Non-Emergency Services: 800/262-8200; 202/887-1315. Operated by the Chemical Manufacturers Association. Handles health and safety questions.

Environmental Defense Fund Recycling Hotline: 212/505-2100; 800/CALLEDF. Information on recycling locations.

Environmental Protection Agency: 800/245-4505; 513/569-7562. Vendors of Innovative Treatment Technologies (VISITT), for vendors that treat groundwater, soil, sludge, sediments, and solid waste.

EPA Emergency Planning and Community Right-to-Know Hotline: 800/334-2467. Information on EPA Title III requirements. 8:30 am to 7:30 pm EST, M-F.

EPA RCRA/Superfund: 800/231-3075. Right-to-Know information for California, Arizona, Hawaii and Nevada. Available 8:30 am to 12 noon, M-F.

EPA User Support: 800/334-2405. EPA mainframe computer user support. Available 8 am to 7 pm, M-F.

National Institutes of Health: 8004-CANCER. Cancer information service.

National Pesticide Telecommunications Network Hotline: 800/858-7378. Information about pesticides; spill handling, disposal, clean-up and health effects. Hotline open 24 hours a day, 365 days a year.

The Pollution Prevention Information Clearinghouse Hotline: 800/424-9346. RCRA/Superfund Hotline: 800/424-9346; Small Business Ombudsman (SBO) Hotline: 800/368-5888; PPIC Technical Assistance: 703/821-4800.

Safe Drinking Water Hotline: 800/426-4791. Provides information on public water supply programs, policy, technical and regulatory items. Hotline available 8:30 am to 5:00 pm EST, M-F.

Solid Waste and Hazardous Waste (RCRA) and Superfund Hotline: 800/424-9346. EPA sponsored: For ordering documents; obtaining regulatory assistance.

Storm Water NPDES Permitting Hotline: 703/821-4823.

Superfund Site Cleanup: 800/533-3508; 214/655-6570. Answers questions regarding status of Superfund sites within Region 6: AR, LA, OK, NM, and TX.

Superfund Technical Information: 800/346-5009. Superfund message center. Allows caller to leave message.

Superfund/RCRA: 800/424-9346; 800/343-3958 or 800/872-2002 or 800/346-5009 or 202/382-3000. Superfund/RCRA Hazardous Waste. Distributes documents and provides regulatory assistance for RCRA hazardous wastes.

Toxic Substances: 800/462-6706; 800/835-6700. Funding for asbestos cleanup projects.

Toxic Substances Control Act Hotline: 202/554-1404. Toxic Substances Control Act (TSCA) & Asbestos Technical Information and Referral.

U.S. Coast Guard, Department of Transportation: 800/424-8802; 202/267-2675. National Response Center for hazardous materials spills.

EPA Whistle Blower Hotline: 800/424-4000; 202/382-3305. For reporting fraud, waste and abuse in EPA programs.

Above sources excerpted from *Environment Today* (November/December 1991), EPA's *Regulatory Assistance for Small Business and Others* brochure (January 1991) and EPA *Memorandum* (update) of January 1993.

-A-

-B-

-C-

-T-

-U-